The Oil Age is Over

**What to Expect as the World Runs
Out of Cheap Oil, 2005-2050**

Matt Savinar, JD

Copyright Notice

About the cover: photo courtesy of fotosearch.com.

ISBN Number: 0-9755118-1-5

Printed in the United States by Morris Publishing
3212 East Highway 30
Kearney, NE 68847
1-800-650-7888

Acknowledgments

This book would not have been possible without the research, writing, bravery, honesty and patience of the following individuals (In no particular order):

Michael C. Ruppert: Editor-in-Chief of *From the Wilderness Publications* (FTW) and author of *Crossing the Rubicon: The Decline of the American Empire at the End of the Age of Oil.*

Dale Allen Pfeiffer: Energy Editor of *From the Wilderness Publications* and author of *The End of the Oil Age.*

Richard Heinberg: author of *The Party's Over: Oil, War, and the Fate of Industrial Civilizations* and *Plan Powerdown: Options and Actions for a Post-Carbon World.*

Julian Darley: author of *High Noon for Natural Gas* and co-founder of *The Post Carbon Institute.*

Dr. Colin J. Campbell: author of *The Coming Oil Crisis* and founder of the Association for the Study of Peak Oil and Gas.

Matthew Simmons: energy investment banker, Simmons and Company and author of numerous white papers on the energy crisis.

Jay Hanson: creator of the Peak Oil and population crash website, http://www.dieoff.org

Dr. Marion King Hubbert (Deceased): for whom the "Hubbert's Peak" is named.

Bob and Anna Savinar: my parents. I wrote the first draft of this book and built my website, http://www.lifeaftertheoilcrash.net, on their computer while living at home. Without their patience neither project would have been completed.

Table of Contents

vi

Part I: Introduction

"Perhaps the sentiments contained in the following pages are not yet sufficiently fashionable to procure them general favor; a long habit of not thinking a thing wrong gives it a superficial appearance of being right, and raises at first a formidable outcry in defense of custom. But the tumult soon subsides. Time makes more converts than reason."
-Thomas Paine, *Common Sense* (1776)

"For my part, whatever anguish of spirit it may cost, I am willing to know the whole truth; to know the worst and provide for it."
-Patrick Henry (1776)

"I am aware that many object to the severity of my language; but is there not cause for severity? I will be as harsh as truth. On this subject I do not wish to think, or speak, or write, with moderation. No! No! Tell a man whose house is on fire to give a moderate alarm; tell him to moderately rescue his wife from the hands of the ravisher; tell the mother to gradually extricate her babe from the fire into which it has fallen -- but urge me not to use moderation in a cause like the present. The apathy of the people is enough to make every statue leap from its pedestal, and to hasten the resurrection of the dead."
-William Lloyd Garrison, *The Liberator* (1831)

"Gas is running low . . ."
-Amelia Earhart (July 2, 1937)

1

Dear Reader,

Civilization as we know it is coming to an end soon. This is not the wacky proclamation of a doomsday cult, apocalypse bible prophecy sect, or conspiracy theory society. Rather, it is the scientific conclusion of the best paid, most widely respected geologists,[1] physicists,[2] and investment bankers[3] in the world. These are rational, professional, conservative individuals who are absolutely terrified by a phenomenon known as global "Peak Oil."

The ramifications of Peak Oil are so serious, one of George W. Bush's energy advisors, billionaire investment banker Matthew Simmons, has acknowledged, "The situation is desperate. This is the world's biggest serious question,"[4] while comparing the crisis to the perfect storm: "If you read *The Perfect Storm*, where a freak storm materializes out of the convergence of three weather systems, our energy crisis results from the same phenomenon."[5]

In May 2001, George W. Bush himself went on the record as saying, "What people need to hear loud and clear is that we're running out of energy in America."[6] In October 2003, Bush's nemesis Michael Moore released the book, *Dude, Where's My Country?* Chapter three of the book, "Oil's Well that Ends Well," was dedicated to the coming post-oil die-off.

If you're like 99 percent of the people reading this letter, you had never heard of the term "Peak Oil" until today. I had not heard of the term until a year ago. Since learning about Peak Oil, I've had my view of the world, and basic assumptions about my own individual future, turned completely upside-down.

A little about myself: In November 2003, I was a 25-year-old law school graduate who found out he had just passed the California Bar Exam. I was excited about a potentially long and prosperous career in the legal profession, getting married, having kids, contributing to my community, and living the "American Dream." Since learning about Peak Oil, those dreams have been radically altered.

I must warn you, the information contained in this book is not for the faint of heart or the easily disturbed. Whether you're 25 or 75, an attorney or an auto mechanic, what you are about to read will likely shake the foundations of your life.

Sincerely,

Matt Savinar

Part II: Peak Oil and the Ramifications for Industrial Civilization

"The world we know is like the Titanic. It is grand, chic, high-powered, and it slips effortless through a frigid sea of icebergs. It does not have enough lifeboats, and those that it has will be poorly employed. If we do not change course, disaster, perhaps catastrophe, is almost inevitable. There is a reason why interest in the Titanic has been revived; it's the perfect metaphor for our planet. On some level we know: we are on the Titanic. We just don't know we've been hit."
-John Brandenburg *Dead Mars, Dying Earth*

"Modern agriculture is the use of land to convert petroleum into food. Without petroleum we will not be able to feed the global population. That is why Professor Watt says, 'We now feel the planet and humanity can only coexist as a living system for a long time if the human population gets down to 1/70 to 4/70 of its current level.'"
-Robert L. Hickerson

"Any number of factors could be cited as the 'causes' of collapse. I believe, however, that the collapse will be strongly correlated with an epidemic of permanent blackouts of high-voltage electric power networks worldwide. Briefly explained: When the electricity goes out, you are back in the Dark Age. And the Stone Age is just around the corner."
-Richard Duncan

"The speed at which a society collapses is directly proportional to the amount of bullshit propagated prior to the collapse."
-Unknown

"This is the first age that's ever paid much attention to the future, which is a little ironic since we may not have one."
-Arthur C. Clarke

"The truth will set you free, but first it will make you sick."
-Unknown

"Your failure to be informed does not make me a wacko."
-John Loeffler

"Deal with reality, or reality will deal with you."
-Dr. Colin Campbell

1.   I heard we have about 40 years of oil left. What's there to worry about?

The statement "we have about 40 years of oil left" is technically correct. The Earth was endowed with about 2,000 billion barrels of oil. We have used about 1,000 billion barrels. As of 2003, we consume about 28 billion barrels per year. 1,000 billion barrels divided by 28 billion barrels per year = 35.7 years of oil left. If one accounts for increased demand resulting from population growth, debt servicing, and further industrialization, that estimate is slashed to a paltry 25 years.

The problem, however, is not "running out of oil" as much as it is "running out of **cheap** oil," which is the resource upon which every aspect of industrial civilization is built. Oil plays such a fundamental role in the world economy that we need not "run out" of the stuff before we run into a crisis of untold proportions.

Like the production of most resources, the production of oil follows a bell curve.[7] The top, or "peak," of the curve coincides with the point at which the respective oil reserve has been 50 percent depleted. The curves of individual oil-fields or oil-producing nations are often asymmetrical due to technological or geopolitical issues, but the global aggregate of these curves comes relatively close to resembling a bell.[8] Regardless of the exact shape of the bell curve, the essential truth is this: oil production goes up, it peaks, then it declines.

The term "Peak Oil" is catchy but slightly misleading in that it suggests a specific date of peak production. In the real world, the top part of the oil production bell curve is almost flat. Once the top of the curve is passed oil extraction becomes increasingly expensive, both financially and energetically.

In practical yet considerably oversimplified terms, this means that if 2000 was the year of global Peak Oil, worldwide oil production in the year 2020 will be the same as it was in 1980. However, the world's population in 2020 will be both much larger (approximately twice) and much more industrialized (oil-dependent) than it was in 1980. Consequently, our need for oil will outstrip our ability to produce it by a huge degree.

The more demand for oil exceeds production of oil, the higher the price goes. The higher the price goes, the more dislocations the world economy suffers. The more dislocations the world economy suffers, the more resource wars the human population endures.

Peak Oil is also known as "Hubbert's Peak," named for Shell geophysicist Dr. Marion King Hubbert. In 1956, Hubbert predicted US domestic oil production would peak around 1970, which it did. He also predicted world oil production would peak around 1995, which it would have had the politically created oil shocks of the 1970s not delayed the world peak for about 10 years. Unfortunately, the day of reckoning is now upon us.

2. When will Peak Oil occur?

The most wildly optimistic estimates indicate 2020-2035 will be the years in which worldwide oil production peaks. Generally, these estimates come from government agencies who openly admit cooking their books, or economists who do not grasp the dynamics of resource depletion. Unfortunately, even in the best-case scenario, petrochemical civilization will begin collapsing by the time today's newborns are old enough to be drafted.

A more realistic estimate is between the years 2004-2010.[9] This is the estimate most frequently given by independent, retired, and now disinterested geologists and former oil industry insiders.[10]

Unfortunately, we won't know we've hit the peak until 4-6 years after the fact. Even on the upslope of the curve, oil production varies a bit from year to year due to factors such as war, weather, and the state of the world economy. It is possible that worldwide oil production peaked in the year 2000 as production of conventional oil has grown only slightly since then. The production of so called "non-conventional" oil may extend the "oil-peak" into what Richard Heinberg calls, "the petroleum-plateau,"[11] that with much luck and prayer, will last until about 2015.

The oil companies have quietly acknowledged the seriousness of the situation. For instance, in a February 1999 speech to oil industry leaders, Arco chairman Mike Bowlin stated, "The last days of the age of oil have begun."[12] Similarly, in a 2003 paper posted on the Exxon-Mobil Exploration website, company president Jon Thompson stated:

> By 2015, we will need to find, develop and produce a volume of new oil and gas that is equal to eight out of every 10 barrels being produced today. In addition, the cost associated with providing this additional oil and gas is expected to be considerably more than what the industry is now spending.

> Equally daunting is the fact that many of the most promising prospects are far from major markets — some in regions that lack even basic infrastructure. Others are in extreme climates, such as the Arctic, that present extraordinary technical challenges.[13]

If Mr. Thompson is that frank in an article posted on the Exxon-Mobil webpage, one wonders what he says behind closed doors.

The Saudis are no less frank than Mr. Thompson when discussing the imminent end of the oil age among themselves. They have a saying that goes, "My father rode a camel. I drive a car. My son flies a jet airplane. His son will ride a camel."[14] As Figure 1 illustrates, the Saudis are not exaggerating:

Figure 1: World Oil Production, 1950-2050
(Source: Dr. C.J. Campbell/Petroconsultants, 1996)

To make matters considerably worse, much of the oil on the bottom half of the down slope may not even be energetically recoverable.

To understand why, you must understand the concept of "net energy." In order to use oil, you have to first search for it, drill for it, extract it, transport it, refine it, and distribute it. Each step requires oil-powered machinery and methods. The ratio between how much energy it takes to acquire an amount of oil and how much energy is contained in the oil is known as "Energy Return on Energy Invested" (EROEI).

Oil used to have an EROEI of about 30 to 1. By the nineties, the ratio had fallen to 5 to 1. Many experts estimate that within a few years, the ratio will be 1 to 1.[15] In other words, it will take a barrel of oil to get a barrel of oil. Once the ratio hits 1 to 1, oil may be technically recoverable, but it will be of no use to us as an energy source. At that point, it won't matter how much money we throw at the process, as oil will cease to be a thermodynamically viable source of energy.

Furthermore, all forms of alternative energy, from solar panels to wind turbines, to nuclear power plants require significant amounts of oil for their initial construction and continued maintenance. Thus, once oil's "EROEI" hits 1 to 1 we will be, for all intents and purposes, out of gas forever.

3. Big deal. If gas prices get high, I'll just carpool or get one of those hybrid cars. Why should I give a damn?

You should give a damn because petrochemicals are key components to much more than just the gas in your car. In short, your entire way of life revolves around the consumption of petrochemicals and fossil fuel energy:

A. Does modern food production depend on oil?

Yes.

As Dale Allen Pfeiffer points out in his FTW article entitled, "Eating Fossil Fuels," approximately 10 calories of fossil fuels are required to produce every 1 calorie of food eaten in the US.[16]

The size of this ratio stems from the fact that every step of modern food production is fossil fuel and petrochemical powered:

1. Pesticides are made from oil;

2. Commercial fertilizers are made from ammonia, which is made from natural gas;

3. Farming implements such as tractors and trailers are constructed and powered using fossil fuels;

4. Food storage systems such as refrigerators usually run on electricity, which most often comes from natural gas or coal;

5. Food distribution networks are entirely dependant on oil. Most of the food at your local super market is packaged in plastic, which comes from petroleum. In the US, the average piece of food is transported almost 1,400 miles before it gets to your plate;[17]

In short, people gobble oil like two-legged SUVs.

Oil-based agriculture is primarily responsible for the world's population exploding from 1.5 billion at the middle of the 19th century to 6.4 billion at the beginning of the 21st. As oil production went up, so did food production. As food production went up, so did the population. As the population went up, the demand for food went up, which increased the demand for oil.

Put simply, the end of cheap oil means end of oil-powered agriculture, which means the end of cheap food, which means the end of billions of lives.

B. Does the delivery of fresh water depend on fossil fuels?

Yes.

Fossil fuels are used to construct and maintain aqueducts, dams, sewers, wells, to desalinate brackish water, and to pump the water that comes out of your faucet. Seven percent of the world's commercial energy consumption is used to deliver fresh water.[18] Most of this energy comes from fossil fuels. Consequently, the cost of fresh water will soar as the cost of oil soars.

C. Does modern medicine depend on oil?

Yes.

Oil is also largely responsible for the advances in medicine that have been made in the last 150 years. Oil allowed for the mass production of pharmaceutical drugs, surgical equipment, and the development of health care infrastructure such as hospitals, ambulances, roads, etc.[19] Consequently, the cost of medical care will soar as the cost of oil soars.

D. Is there anything in the modern world that doesn't depend on oil?

No.

In addition to transportation, food, water, and modern medicine, mass quantities of oil are required for all plastics, the manufacturing of computers and communications devices, extraction of key resources such as copper, silver, and platinum, and even the research, development, and construction of alternative energy sources like solar panels, windmills, and nuclear power plants.

E. Conclusion

The aftermath of Peak Oil will extend far beyond how much you will pay for gas. If you are focusing solely on the price at the pump and/or more fuel-efficient forms of transportation, you aren't seeing the bigger picture. Converting your car to run on biodiesel won't do you much good if there isn't enough energy to maintain roads and highways. Purchasing a hybrid car will seem a bit pointless when you don't have a job to drive to because the economy has collapsed due to oil depletion. Spending $10,000 to install solar panels on your roof won't provide you with much comfort when our fossil fuel powered food and water distribution infrastructure has ceased to function.

In short, the end of cheap oil means the end of everything you have grown accustomed to, all aspects of industrial civilization, and quite possibly humanity itself. This is known as the post-oil "die-off."

4.   What did you mean by "die-off"?

Exactly what it sounds like. It is estimated the world's population will contract to less than 500 million within the next 50-100 years as a result of oil depletion (current world population: 6.4 billion).

5.   Are you serious? That's 90 percent of our current population. How could that many people perish? Where does that estimate come from?

That estimate comes from biologists who have studied what happens to every species when it overshoots its resource base. When given access to an abundance of food (energy) all species follow the same pattern: a rapid population increase ("overshoot") followed by an even more rapid population decrease characterized by violence, war, and cannibalism.

Example A: Bacteria in a Petri dish

Bacteria in a Petri dish will grow exponentially until they run out of resources, at which point their population will crash. Only one generation prior to the crash, the bacteria will have used up half the resources available to them. To the bacteria, there will be no hint of a problem until they starve to death.

But humans are smarter than bacteria, right? You would think so, but the facts seem to indicate otherwise. The first commercial oil well was drilled in 1859. At that time, the world's population was about 1.5 billion. Less than 150 years later, our population has exploded to 6.4 billion. In that time, we have used up about half the world's recoverable oil. Of the half that's left, most will be very expensive to extract. If the experts are correct, we are less than one generation away from a crash. Yet to most of us, there appears to be no hint of a problem. One generation away from our demise, we are as clueless as bacteria in a Petri dish.

Example B: Reindeer on St. Matthew Island

In 1944, researchers moved a population of 29 reindeer to St. Matthew Island, an unoccupied island in the Bering Sea. Luckily for the reindeer, the island had an abundant supply of their favorite food: lichen. With food readily available, the reindeer population exploded to 6,000 by 1963. At that point, reindeer were everywhere to be seen on the island. By 1966, however, the only things to be seen on the island were reindeer skeletons. In those three years, the reindeer had consumed all of the island's lichen. As a result, the reindeer population crashed to a total of 42.[20]

Take a look at Figure 2, which charts the reindeer population on St. Matthew Island from 1944 to 1966. Compare the shape of Figure 2 with the shape of Figure 3, which charts the (projected) human population on Earth from the year 1850 to the year 2050. You will notice that both charts follow a "J" curve. With access to an abundant (fossil fuel-powered) food supply, our population has grown just as the reindeer population grew when it had access to an abundant food source.

The reindeer on St. Matthew Island relied almost exclusively on the island's lichen supply to sustain themselves, in much the same way that we rely almost exclusively on fossil fuels to sustain ourselves. When the Earth's supply of readily available fossil fuels runs out, the result for us will likely be the same as it was for the reindeer when the island's supply of readily available lichen ran out.

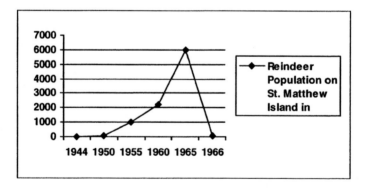

Figure 2: Data Adapted from "The Introduction, Increase, and Crash of Reindeer on St. Matthew Island." Source: Dr. David R. Klein, Cooperative Wildlife Research Unit, University of Alaska.

12

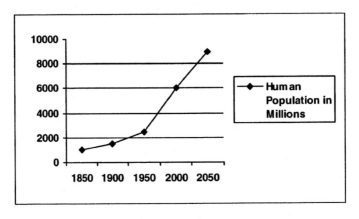

Figure 3: Human Population 1850-2050

I know what you're thinking: "But we're human beings with higher consciousness. Certainly we can figure out a way to avoid the fate suffered by a group of reindeer." Unfortunately, as the next example demonstrates, we appear to be repeating the same mistakes of previous human populations who suffered through die-offs.

Example C: Easter Island

Over the course of human history, many populations have suffered from die-offs. In fact, large-scale regional die-offs of human populations are quite common and very well documented. The only thing uncommon about the impending die-off is its scope: it will be a global die-off, not a regional one.

Of all previous die-offs, the one most analogous to our current situation is the die-off that took place on Easter Island beginning in the late-1500s.

Easter Island was discovered by Western civilization in 1722 when Dutch explorer Jacob Roggeveen landed on the island. At the time, Roggeveen described the island as, "a singular poverty and barrenness."[21] As UCLA Medical Professor Jared Diamond explains, the islanders Roggeveen encountered had no wheels, no firewood, no farming implements, no animals larger than insects, and only a handful of flimsy, leaky canoes.[22]

The primitive lifestyle led by the islanders stood in stark contrast to the giant, elaborately constructed, stone statues that littered the island. Roggeveen and his crew were completely perplexed by these statues, as it was clear whoever built them had tools, resources, and organizational skills far more advanced than the islanders they encountered.

What happened to these people?

According to archeologists, Easter Island was first colonized by Polynesians sometime around the year 500 AD. Over the next 1,000 years, the island's population grew to anywhere from 7,000-20,000 depending on who's making the estimate.[23] During this population boom, the islanders used wood from the forest trees to power virtually every aspect of a highly complex society. Professor Diamond writes, "The people used the land for gardens and the wood for fuel, canoes, and houses-and, of course, for lugging statues."[24]

Eventually, the islanders began cutting the trees down faster than the trees could grow back. Amazingly, there is evidence that the islanders actually intensified their statue building efforts as the supply of timber dwindled. This has led archaeologists to conclude that the islanders never bothered to figure out how much timber they had in "reserve."[25] Of course, another possibility is that whoever the islanders selected to figure out how much timber they had in "reserve" was either lying or incompetent.

In either case, the supply of timber went into terminal decline. Professor Diamond describes what happened next:

> . . . the islanders ran out of timber and rope to transport and erect their statues. Life became more uncomfortable: springs and streams dried up, and wood was no longer available for fires . . . chaos replaced centralized government and a warrior class took over from the hereditary chiefs. People took to living in caves for protection against their enemies.[26]

The chaos became so widespread the island is still littered with the remnants of weapons used by the islanders.[27] Food became so scarce the islanders resorted to cannibalism. The practice became so common the islanders developed a new insult, "The flesh of your mother sticks between my teeth."[28]

If you've got a sick feeling in your stomach right about now, there is good reason. As Professor Diamond notes, "Easter Island looks like a perfect metaphor for us."[29]

Like the islanders, we have built our entire civilization around one resource. Our entire culture and psychology revolves around the distribution and consumption of that resource. That resource drives our transportation, housing, and food and water distribution networks. We have no true alternatives to that resource. As the supply of that resource began to dwindle, a highly militarized cabal began dismantling the long-standing and highly respected government of the world's most powerful nation. This cabal then plunged the world into ever-intensifying wars over access to that resource.

14

If the similarities between the initial stages of collapse on Easter Island and the ecological and geopolitical events of the past few years are undeniable, the future implications of these similarities are nothing short of mind blowing. On Easter Island, the children of the survivors lost all the knowledge and technological abilities of their ancestors. They were clueless as to the construction or operation of the sea faring vessels their ancestors had traveled to the island in. When Roggeveen's crew inquired as to how the statues had been built, the islanders were unable to offer an explanation other than the statues had "walked" across the island.[30]

As far as our descendants are concerned, the ramifications of a similarly drastic loss of knowledge are stunning. It is entirely possible our descendents (if there are any) will be clueless as to how the empty skyscrapers and abandoned automobiles they see were originally constructed.

Even if they possess a somewhat accurate understanding of how the skyscrapers and automobiles were created, they are unlikely to know how to operate an elevator, a light switch, or a car ignition. With so little energy available to operate these devices, there will be little reason for anybody to learn how.

Eventually, the idea that man once visited the Moon will, quite possibly, be considered a fairy tale. Our descendants may look at pictures of NASA space missions in much the same way we look at cave paintings. For all we know, they may come to believe Neil Armstrong and Buzz Aldrin were the fictional cohorts of James T. Kirk and Jean Luc-Picard.

6. I still can't imagine that number of deaths. It's just too ghastly to imagine. How can that possibly be?

I know how you feel. This is all very difficult to handle, both emotionally and intellectually. The implications are quite staggering. Perhaps the following explanation, while considerably over-simplified, will help illustrate the future we are marching towards.

As explained above, worldwide oil production follows a bell curve. Thus, if the year 2000 was the year of peak production, then oil production in the year 2025 will be about the same as it was in the year 1975. The population in the year 2025 is projected to be roughly 8 billion. The population in 1975 was roughly 4 billion. Since oil production essentially equals food production and distribution, this means that we will have 8 billion people on the planet but only enough food/fuel for 4 billion. The further we go into the future, the less oil we will have to produce and distribute food. So the ratio of people to oil/food will just keep getting worse and worse.

15

With that in mind, visualize the following situation: you, me, and six other people are locked in a room, with only enough food for four of us.

The math is simple: at least four of us will die from starvation; another one or two will likely die as we all fight each other for what little food we have.

That's what will happen if we're just fighting with our fists. Give each of us weapons and you can imagine what that room is going to look like by the time we're finished with each other.

7.   Is it possible that it may now be too late to stop the die-off?

As much as I hate to admit it, yes.

Personally, I refuse to accept the die-off as inevitable. I am the first to admit, however, that my refusal is based on hope and faith, not facts and science. As biologist David Price explains in his 1995 article, "Energy and Human Evolution:"

> A population that grows in response to abundant but finite resources tends to exhaust these resources completely. By the time individuals discover that remaining resources will not be adequate for the next generation, the next generation has already been born. And in its struggle to survive, the last generation uses up every scrap, so that nothing remains that would sustain even a small population.[31]

Unfortunately, the parallels between the populations Price refers to and the human populations are impossible to dismiss. Only recently have more than a handful of us realized we don't have enough oil to last for more than another generation. Even fewer of us have realized none of the alternatives to oil, or combination thereof, can deliver more than a small fraction of the energy required by industrial civilization. If the last 4-5 years are any indicator of what is to come, we will spend the next generation fighting for every last drop of the stuff.

16

8. Not to be insensitive, but I'm sure that most of those deaths will take place in the third world. The US will likely have the bulk of the survivors, right?

Not by a long shot.

In their 1994 article entitled, "Food, Land, Population, and the US Economy," researchers David Pimentel and Mario Giampetro make the following points:

1. The population in the US is increasing at a rate of 1.1 percent per year, not including illegal immigration. At this rate, the US population will reach 520 million by the year 2050.

2. As urbanization and soil erosion continue unabated, the US is projected to only have 290 million acres of arable land by 2050. With a population of 520 million, that means each person will only have .6 acres of arable land from which they can derive their food. Agronomists stress, however, that a person needs a minimum of 1.2 acres of arable land for a productive diet.

3. Americans currently consume approximately 1,500 gallons of water per day/per person to meet all their needs. (This includes industry, transportation, national defense, food production, etc., not just the water you drink individually.) Hydrologists estimate that a human needs a minimum of 700 gallons of water per day/per person to meet their basic needs. At our current rate of population growth, we will only be able to deliver 700 gallons per day/per person by 2050 — just barely enough water for each person. [32]

As terrifying as Pimentel and Giampetro's calculations are, the "real-life" scenario will be even worse for two reasons: their calculations don't account for the devastating impact the coming oil shocks will have on agricultural productivity or the fact the US has hamstrung its ability to address these issues through troublesome international trade entanglements.

Our economic and industrial entanglements with China, for instance, will prove particularly problematic. China's 100-billion-dollar trade surplus with the US gives it tremendous power to affect the US economy. As China's food shortage becomes more severe, China will threaten to pull the plug on the US economy if the US does not provide China with food. At that point, however, the US will itself be in the midst of a severe food shortage. If it gives into China's demands, US citizens will starve. If it refuses to give into China's demands, China will withdraw its investments from the US economy. A massive economic meltdown will then ensue, with many US citizens going hungry as a result.

If you find the possibility of mass starvation in the US hard to believe, keep in mind the average piece of food travels nearly 1,400 miles before it gets to your plate. Very little of the food you eat is produced in your local community.

As fuel prices skyrocket, transportation systems will begin to breakdown. The delivery of food from 1,400 miles away will become a pragmatic and financial impossibility.

Large cities like Los Angeles, New York, San Francisco, and Miami will become truly horrific places to live as you can't grow crops on concrete. Without affordable fuel, little food or water will get in and few people will be able to get out. To make matters worse, the high cost of fuel will impede the ability of police, fire, and medical services to maintain order.

The bottom line is this: if we don't start taking these issues seriously, our status as Americans will do nothing to save us from the fate of the rest of the world.

9. Clearly, we have a real problem, but you're describing the worst-case scenario, right?

I'm describing the most likely scenario. The worst-case scenario is extinction, as the wars that will accompany the worldwide oil shortage will likely be the most horrific and widespread humanity has ever experienced. We will discuss this more in Part VII.

10. What do Dick Cheney and George W. Bush have to say about this?

In late 1999, Dick Cheney stated:

> By some estimates, there will be an average of two-percent annual growth in global oil demand over the years ahead, along with, conservatively, a three-percent natural decline in production from existing reserves. That means by 2010 we will need on the order of an additional 50 million barrels a day.[33]

To put Cheney's statement in perspective, remember that the oil producing nations of the world are currently pumping at full capacity but are unable to produce much more than 80 million barrels per day. Cheney's statement was a tacit admission of the severity and imminence of Peak Oil as the possibility of the world raising its production by such a huge amount is borderline ridiculous.

A report commissioned by Cheney and released in April 2001 was no less disturbing:

> The most significant difference between now and a decade ago is the extraordinarily rapid erosion of spare capacities at critical segments of energy chains. Today, shortfalls appear to be endemic. Among the most extraordinary of these losses of spare capacity is in the oil arena.[34]

Not surprisingly, George W. Bush has echoed Dick Cheney's sentiments. In May 2001, Bush stated, "What people need to hear loud and clear is that we're running out of energy in America."[35]

One of George W. Bush's energy advisors, energy investment banker Matthew Simmons, has spoken at length about the impending crisis. If you have a diehard conservative friend or relative who insists talk of an "oil crash" is nothing but eco-fascist fear mongering, you might want to direct them to the archive of Simmons' speeches and papers available online at http://www.simmonsco-intl.com or the archive of Simmons' audio and video interviews available online at http://www.globalpublicmedia.com.

Simmons is a self-described "lifelong Republican." His investment bank, Simmons and Company International, is considered the most reputable and reliable energy investment bank in the world.

Given Simmons' background, what he has to say about the situation is truly terrifying. For instance, in an August 2003 interview with *From the Wilderness* publisher Michael Ruppert, Simmons was asked if it was time for Peak Oil to become part of the public policy debate. He responded:

> It is past time. As I have said, the experts and politicians have no *Plan B* to fall back on. If energy peaks, particularly while 5 of the world's 6.5 billion people have little or no use of modern energy, it will be a tremendous jolt to our economic well-being and to our health — greater than anyone could ever imagine.

When asked if there is a solution to the impending natural gas crisis, Simmons responded:

> I don't think there is one. The solution is to pray. Under the best of circumstances, if all prayers are answered there will be no crisis for maybe two years. After that it's a certainty.[36]

19

In May 2004, Simmons explained that in order for demand to be appropriately controlled, the price of oil would have to reach $182 per barrel. With oil prices at $182 per barrel, gas prices would likely rise to $7.00 per gallon.[37]

If you want to ponder just how devastating oil prices in the $180 range will be for the US economy, consider the fact that one of Osama Bin-Laden's goals has been to force oil prices into the $200 range.[38]

11. Is it possible we are already in the first stages of the crash?

Yes. Ample evidence exists that we are already crashing:

A. Has world average oil and energy production per-capita already peaked and been in terminal decline for almost an entire generation?

Yes.

In an article entitled "The Peak of World Oil Production and the Road to the Olduvai Gorge," geologist Dr. Richard Duncan makes the following points:

1. The amount of oil available per person grew from one half barrel per year in 1920 to a peak of 5.5 barrels per year in 1979. It has since declined to 4.32 barrels per person per year.

2. The amount of total energy from all sources (oil, coal, natural gas, nuclear, renewables) available per person from all sources also peaked in 1979, and has declined by an average rate of .33 percent per year since then. The rate of decrease will accelerate to an average of 5.45 percent per year beginning in 2012. At that point, the electrical grid will begin rapidly disintegrating. By 2030, industrial civilization will cease to exist.[39]

In laymen's terms, the energy pie has kept getting bigger, but the amount of pie per person has kept getting smaller. Unfortunately, the amount of pie per person is going to continue to shrink at an increasingly rapid rate in the years to come.

This decline in per-capita availability of energy has far reaching effects as the availability of oil is at the base of everything we do. It permeates through all of our political, social, and economic institutions and interactions.

For instance, as the per-capita availability of energy has declined, our ability to provide education and health care on a global scale has also declined

20

while the incidence of global conflict has increased. This is no coincidence, as the construction and maintenance of schools and hospitals requires tremendous amounts of energy while wars most often result from resource shortages, whether real or perceived.

B. Has production of conventional oil flattened in recent years?

Yes. Production of conventional oil has grown only marginally since 2000. Remember, US domestic oil production peaked in 1970. This fact was vehemently denied and constantly explained away until about 1975 when it became impossible to hide the truth.

C. Has the rate of decline for some oil producing countries been increasing at an alarming rate in the past couple of years?

Yes.

According to Chris Skrebowski in his recent article for *Petroleum Review* entitled, "Over a Million Barrels of Oil Lost Per Day Due to Depletion:"

1. Eighteen oil-producing countries account for approximately 30 percent of the world's oil production. These 18 countries saw their production decline by an alarming 1.14 million barrels per day during 2003.[40]

2. In 1998 the total production of these 18 countries dropped by less than one percent, whereas last year it declined by nearly five percent.[41]

The rate at which oil production in these countries is not only dropping, it is dropping at an increasingly rapid (and alarming) rate.

D. Have estimates of oil & natural gas reserves been revised drastically downwards in recent years?

Yes.

In October 2003, *CNN International* reported that a research team from Sweden's University of Uppsala had discovered worldwide oil reserves are as much as 80 percent less than previously thought, that worldwide oil production will peak within the next 10 years, and once production peaks, gas prices will reach disastrous levels.[42]

In January 2004, shares of major oil companies fell after Royal Dutch/Shell Group shocked investors by slashing its "proven" reserves 20 percent, raising concerns others may also have improperly booked reserves.[43]

In February 2004, energy company El Paso Corporation announced it had cut its proven natural gas reserves estimate by 41 percent.[44] Between February and May 2004, Shell downgraded its reserves three more times.[45] This completely rocked the oil industry as Shell's reputation for honest reporting was above and beyond that of any other major oil company.

By June 2004, Shell executives were so beleaguered by various scandals they openly admitted their exploration and production activities had "inadvertently" fed poverty, corruption, and violence throughout the world.[46] Later that month, the CEO of Shell, Sir Watts, was fired and granted a 1.8 million dollar buyout.[47]

Two weeks prior to Watts' buyout, the investment bank Goldman Sachs announced oil giant British Petroleum might also be downgrading its reserves.[48]

E. Have oil and gas prices skyrocketed in recent years?

Yes.

In 1998-1999, the price of oil hovered around $11.00 per barrel. By August 2004, the price had risen over 450 percent, nearly touching $50 per barrel.

In 1998-1999, the average price of gas in the US hovered around $1.00 per gallon. By June 2004, the average prices of gas in the US had doubled to an all-time record $2.00 per gallon. In San Francisco, gas surpassed $2.50 per gallon throughout June 2004, while a few remote localities in Northern California area actually flirted with the $3.00-per-gallon mark.

F. Has spare capacity evaporated to virtually nothing?

Yes.

"Spare capacity" is defined as how much extra oil production can be brought online within 30 days notice and maintained for 90 days thereafter. As Amy Jaffe of the Baker Institute has explained, over the past 20 years OPEC's spare capacity has dwindled to almost nothing:

> In 1985 OPEC maintained about 15 million barrels per day of spare capacity — about 25 percent of world demand at that time.

> In 1990, OPEC maintained 5.5 million barrels per day of spare capacity — about 8 percent of world demand.

By 2003, OPEC's spare capacity had shrank to 2 million barrels per day — about 2-3 percent of demand.[49]

In the summer of 2004, the president of OPEC announced, "The oil price is very high. It's crazy. **There is no additional supply.**"[50] (Emphasis added)

When the president of OPEC says the price of oil is "very-high" and "crazy," it's safe to say the era of cheap oil is over.

A few weeks later, the Energy Information Agency announced production was at 99 percent of capacity.[51] In other words, spare capacity had temporarily vanished altogether.

G. Has large-scale civil unrest broken out as a result of high fuel and food prices?

Yes.

In fall of 2000, Israel, France, Spain, the UK, and the Netherlands were besieged by large-scale gas price protests. The UK government even threatened to use force to halt the protests.

In June 2001, police in Indonesia fired tear gas at thousands of protesters angry at fuel price increases.[52]

In October 2003, gas protests in Bolivia resulted in 70 deaths and 400 injuries.[53]

In April 2004, truckers in Los Angeles protested near record fuel prices by parking their rigs on a busy freeway during rush hour traffic.[54] One month later, truckers in California staged protests at the Port of Oakland to demand higher pay so they could handle crushing fuel costs.[55]

In June 2004, the British government announced that hundreds of troops would be deployed to defend vital supermarket depots in the event of fresh fuel protests during the fall.[56]

Keep in mind the British people are nowhere near as angry, armed, or accustomed to cheap fuel and food as their US counterparts. One can only imagine what measures the US government will resort to in order to control its population as food and fuel shortages begin to hit hard.

H. Have marauding gangs of "Mad Max"-style fuel bandits formed yet?

Yes.

Fuel-siphoning crime syndicates have already formed in Central California. These syndicates roam the countryside at night in search of diesel fuel stored on the land of small farmers. The syndicates steal the fuel and sell it on the black market. *The New York Times* quoted one farmer victimized by the syndicates as stating, "I don't come to the door without a .357 magnum in hand."[57]

If you think that's disturbing, consider the fact China now executes people for stealing fuel.[58]

I. Does the sluggish economy have anything to do with declining oil production?

Yes.

You can think of "Peak Oil Production" as a synonym for "Peak Job Creation." As of August 2004, the government is still insisting the unemployment rate is in the 6 percent range. This seemingly low number is misleading as it does not account for the fact so many of the jobs created in the last few years have been either part-time or low-paying.

The unemployment numbers are so "cooked" by the government these days that they are essentially meaningless. Nobody can say for sure what the "real" unemployment rate is, but I've seen numerous well-reasoned and logical estimates that place it anywhere from 9-15 percent.

In the United States, we need to create over 250,000 new jobs per month just to keep up with population growth. Creating new jobs is essentially impossible now that oil production is peaking. Without an excess supply of energy, the economy cannot grow, and the necessary number of full-time, living-wage jobs cannot be consistently created. Once oil production begins to decline, it will be impossible to stop the economy from hemorrhaging more jobs with each passing year. If you are not employed come Peak Oil, your chances of finding a job will diminish with each passing year. If you are employed come Peak Oil, your chances of losing your job or receiving a pay cut will grow with each passing year.

With fewer people employed, the nation's tax base will diminish. As a result, local, state, and national governments will have less money with which to fund everything from public schooling to police and fire services to subsidies for alternative energy.

J. *Have large-scale blackouts become a common occurrence in recent years?*

Yes.

The rolling blackouts experienced in California during fall of 2000, the massive East Coast blackout of August 2003, and the various other massive blackouts that occurred throughout the world during late summer of 2003 are simply a sign of things to come. The report commissioned by Dick Cheney referenced in question 10, for instance, warned that the US can expect "more Californias."

On August 10, 2004, *The Washington Post* echoed the report's prediction when it explained "blackouts are now inevitable," and rather than try to avoid them, we should do our best to cope with them.[59]

Large-scale blackouts are now a regular occurrence in many parts of the world, particularly China, the US's chief energy rival. From January to April 2004, 24 of China's 31 provinces were hit by power cuts and partial blackouts or "brownouts."[60] In July 2004, Beijing's grid switched off power to parts of 10 districts in the capital after several electricity generators broke down.[61] The energy crisis in China has become so severe the government has asked businessmen to stop wearing suits so they don't have to use air conditioning.[62]

Massive blackouts have also hit Greece, Singapore, France, England, and Spain.

K. *Have food and chemical production been declining in recent years?*

Yes.

World grain production has dropped every year since 1996-1997.[63] World wheat production has dropped every year since 1997-1998. Recent food price hikes in China are probably just the beginning of a massive international food crisis.[64] The natural gas crisis of 2003 forced many US fertilizer factories to shut down or slow their production.[65] By 2004, the rising price of agricultural energy inputs was costing US farmers $6 billion dollars in added expenses per year,[66] while the chemical industry was experiencing major difficulties keeping prices under control as the result of the rising cost of natural gas and other petrochemical feedstock.

12. This is going to be a slow, gradual decline, right?

Probably not.

According to conservative estimates, once we pass the peak, oil production will decline by 1.5-3 percent per year. However, the decline is more likely to be over 5 percent per year as the rising price of oil will motivate oil companies to frantically drill for whatever of the black stuff is left, thereby pushing production past the plateau and off the cliff.

Many countries are seeing their oil production drop at an even faster rate. In a recent article for *Petroleum Review*, Chris Skrebowski explained that during 2003:

1. Gabon, whose production peaked in 1996, saw its production drop by an alarming 18 percent;

2. Australia saw its production drop more than 14 percent;

3. UK production from the North Sea declined by 9 percent;

4. Indonesia (an OPEC country) saw its production drop by 8.5 percent. [67]

If the rest of the world declines at a rate comparable to these nations, a drop in global oil production of 40-60 percent within 10 years of the peak is not completely out of the realm of possibility. If we're extraordinarily lucky, and all current trends are bucked, production may drop by only 25 percent in the 10 years following the peak. This is still an absolutely huge amount given the importance of oil to the world economy.

The effects of the physical drop in production will almost certainly be exacerbated by disruptions in supply resulting from war and terrorism, as an increasing percentage of the world's oil supply will be coming from unstable countries like Iraq, Iran, Saudi Arabia, and Nigeria.

13. What about the oil in the Arctic National Wildlife Preserve (ANWR)? Can't we drill there?

At current rates of oil consumption, the ANWR contains enough oil to power the US for only six months.[68] The Energy Information Administration has estimated tapping ANWR would lower oil prices by a whopping 50 cents per barrel.[69] The fact that it is being touted as a "huge" source of oil underscores how serious our problem really is.

14. What about the oil under the Caspian Sea?

Prior to the war on terror, the Caspian Sea was thought to contain over 200 billion barrels of oil. Shortly after invading Afghanistan, the US discovered (much to its dismay) the Caspian Sea probably only holds between 20 and 40 billion barrels.[70]

While this is far from an insignificant amount of oil, its value is largely offset by the neighborhood it's located in. The area around the Caspian Sea has the potential for wars and disruptions that could make the current debacle in Iraq look like a late night jaunt through New York's Central Park. That's not really much of an exaggeration when you consider the countries surrounding the Caspian Sea: Russia, Kazakhstan, Turkmenistan, Uzbekistan, Iran, and Azerbaijan.

15. What about so-called "non-conventional" sources of oil?

So called "non-conventional" oil, such as the oil sands found in Canada and Venezuela, is incapable of replacing conventional oil for the following reasons:

1. The EROEI of non-conventional oil is very poor compared to conventional oil, clocking in at about 3/2.[71] The cost of oil from the Canadian oil sands projects became so high in May 2003 that the oil industry publication *Rigzone* suggested, "President Bush, known for his religious faith, should be praying nightly that Petro-Canada and other oil sands players find ways to cut their costs and boost US energy security."[72]

On a similar, albeit less religious note, Shell-Canada Senior Vice President Neil Camarta has explained the oil sands of Canada are much more expensive than free-flowing crude of the Persian Gulf, "It's not like the oil in Saudi Arabia. You see all the work we have to do; it doesn't just jump out of the ground. Every step takes brute force."[73]

If Camarta is under no illusion that the oil sands can replace conventional oil, you shouldn't be either. Brute force requires lots of excess energy, which is exactly what we will have less of as we slide down the down-slope of the conventional oil production curve.

2. The environmental costs of the oil sands projects are absolutely staggering and the process uses a tremendous amount of fresh water and natural gas, both of which are in limited supply. The oil sands projects in Canada have consumed so much natural gas that Alberta has even considered installing a nuclear power plant in the middle of the project to supply power and steam.[74]

3. Although non-conventional oil is quite abundant, its rate of extraction is so slow it will, at best, only slightly ameliorate the coming crisis. According to Dr. Colin Campbell, combined Canadian and Venezuelan output of non-conventional oil will reach only 4.6 mbd by 2020.[75] Unfortunately, once the decline really gets under way, we will be losing over 1.5 million barrels per day/per year due to depletion. Thus, even in the best-case scenario, the oil sands only buy us an extra three years or so before the decline begins in earnest.

On a positive note, the production of non-conventional oil may push the peak for total oil production back by about 4 years from 2004-2008 to about 2008-2012. This gives those of us who are "Peak-Oil informed" a bit more time to prepare.

16. I just read an article that states that known oil reserves keep growing.

There is probably good reason to doubt the veracity of recent reports of "reserve growth." Most reserve growth can be attributed to one of three factors:

1. In recent years, the USGS and the EIA have revised their estimates of oil reserves upwards. Peak Oil "deniers" often point to these revisions as proof that fears of a global oil shortage are unfounded. Unfortunately, these upwards revisions are highly suspicious. For instance, after recently revising oil supply projections upward, the EIA stated, "These adjustments to the estimates are based on non-technical considerations that support domestic supply growth to the levels necessary to meet projected demand levels."[76] In other words, they predicted how much they think we're going to use, and then told

us, "Guess what, nothing to worry about — that is how much we've got!"

The track record of the USGS isn't much better than that of the EIA. In March 2000, for instance, they released a report indicating more "reserve growth." Colin Campbell responded to the report as follows:

> Let us not forget that McKelvey, a previous director of the USGS, succumbed to government pressure in the 1960's to discredit Hubbert's study of depletion, which was subsequently vindicated in the early 1970's after US production actually peaked as Hubbert had predicted. It did so by assuming that all the world's basins would be as prolific as Texas in a very damaging report . . . that successfully misled many economists and planners for years to come.[77]

Similarly, Richard Heinberg reminds us, "in 1973, Congress demanded an investigation of the USGS for its failure to foresee the 1970 US oil production peak."[78]

2. During the late 1980s, several OPEC countries drastically increased their reported oil reserves even though they had no major oil discoveries. How is that possible?

The answer probably has something to do with the fact an individual OPEC member's quotas are proportional to their proven reserves. The more they report in reserves, the more they are allowed to export, which means the more money they make. Thus, they have a huge incentive to report "reserve growth."

3. The oil sands in Canada were recently reclassified in a fashion that boosted the world's total reported oil reserves by a considerable amount. As explained previously, the oil sands are abundant, but are plagued by issues of high cost and slow extraction rate. Thus, the reserve growth resulting from the reclassification of the oil sands is nothing to get too terribly excited about.

Oil companies, politicians, and government agencies all have massive psychological and financial incentives to accept these upward estimates because our economy is based on growth. If a report comes out saying oil reserves have grown, it is unlikely to be questioned.

Finally, it is important to note that the bottom line for the world's population is how much oil can be produced and at what price. It doesn't matter if the world has X amount in reserve if it can't produce enough per day to allow for continued economic growth.

When some expert tells you, "No need to be concerned. We have X amount in reserve. With so much in reserve, only doomsayers claim a crisis is at hand," ask him, "Fine, but how much can we **produce** per day?" Right now, the world needs over 82.5 million barrels of oil per day to sustain economic growth. If the world is only able to produce 70, 60, 50, or 40 million barrels per day, it matters little how much we have in reserve.

This is a relatively easy concept to understand for somebody like yourself who actually reads books. However, since you may have a friend or relative who you need to convince via the spoken word, here is an analogy that might be useful when comes time to explain why the large reserve numbers cited by the "obfuscators" are misleading:

Pretend you have 10 children and can barely afford to feed and clothe all of them. One day, you inherit a $1,000,000 dollar bank account from an anonymous long-lost relative.

You are all set to spend to your heart's delight until your attorney explains the fine print of the will to you. It turns out the will stipulates you can only withdraw $80 per day during the first year, $78 per day during the second year, $75 per day during the third year, $74 per day during the fourth year, $71 per day during the fifth year and so on until the account is drained.

In other words, you have $1,000,000 "in reserve," during the first year but you can only "produce" $29,200 per year. That's not going to go very far considering you've got 10 kids to feed and tons of debt to service. By the fifth year, you can only produce $25,550 per year. The amount you can produce will continue to shrink with each passing year even though your child-rearing expenses will continue to rise.

Still think you should get that Hummer or does a bicycle make more sense for you and your children?

17. Is it possible there is still more oil left to be discovered?

Almost certainly not. All available evidence indicates that we have already located the overwhelming majority of the world's oil reserves:

1. World oil discovery peaked in 1962 and has declined to virtually nothing in recent years.

2. According to a January 2002 study by the Colorado School of Mines entitled, "The World's Giant Oil Fields," the fourteen largest fields account for over 20 percent of the world's crude oil supply. The average age of these fields is 43.5 years.[79]

    The study explains that the "giant" fields discovered today are tiny compared to the "giant" fields of yesteryear. The study concludes, "Most of the world's true giants were found decades ago. In the past two decades, most oil and gas discoveries have been quite small fields."[80]

3. According to Colin Campbell and Jean Laherrere's 1998 article in *Scientific American*, "The End of Cheap Oil," about 80 percent of oil produced today comes from fields that were discovered more than 30 years ago and most of these fields are well past their peaks.[81]

4. According to a January 2004 report in *Petroleum Review,* the discovery rate for mega projects (those which contain more than 500 million barrels or about a one week supply) has now dwindled to nothing. In 2000, 16 mega projects were discovered. In 2001 there were 8 new discoveries. In 2002 there were 3 new discoveries. In 2003, there were none. [82]

5. According to a January 7, 2004 article in *The Financial Times,* between 2001 and 2003, oil companies discovered less than half the reserves they found between 1998 and 2000.[83]

In fact, discovering oil has become such a challenge that there is now a reality show about it. PBS's "Extreme Oil," documents the "extreme" lengths the oil industry now resorts to in hopes of pulling out whatever affordable oil the Earth has left.

If a reality show about discovering oil isn't a sign the end is near, I don't know what is.

18. What if there is actually a huge amount of oil just waiting to be discovered that we somehow missed? Wouldn't that make a difference?

Not really.

University of Colorado physicist Albert Bartlett has explained that because of the compounding nature of yearly increases in demand, a doubling of the world's original endowment of oil only puts the peak off by 25 years.[84]

In laymen's terms Bartlett is saying the following: we think we originally had about 2,000 billion barrels. We've used up about 1,000 of those, thus putting us right around the peak. Now let's say it turns out that there were originally 4,000 billion barrels. This would leave us with 3,000 billion barrels in the ground instead of the 1,000 billion. However, because of exponential growth in demand, we'd still be within one generation of the peak.

In other words, even if we have three times the amount of oil remaining in the ground as we think we have, we're still going to face a major crisis within most of our lifetimes.

19. But I just heard on the news that they made a huge oil discovery somewhere!

Nowadays, an oil discovery of 200 million barrels is considered huge, and will garner much attention in the press. Such a find will often be cited in the media as "proof" there is no massive oil crisis looming. Such proclamations are designed to placate the average person, who doesn't possess the necessary background information to evaluate what the newscaster is telling them.

The world uses 80 million barrels per day. So even a huge find these days is really only a two- to three-day supply. The discovery will certainly make a lot of money for whoever found it, but it won't do much to soften the coming oil shocks for you and me.

20. Is it possible that things might get better before they get worse?

Yes.

Once an oil find is made, it takes about 5-7 years for production to come online. As stated previously, the last remotely decent year for oil finds was 2000. This means the last decent year for new production to come online will be about 2007. By 2010, those projects will be in decline.

21. Can't technology just find better, more efficient ways to use the oil that is left?

Absolutely, but that increased efficiency means nothing unless we use less oil overall. Thus far, the more energy-efficient we get, the more energy we consume. Given the growth-obsessed nature of our economy, the psychological profile of most Americans, and the corruptness and incompetence of our leadership, that trend is unlikely to change.

The idea that technologically derived increases in energy efficiency will solve this for us is fundamentally flawed: technology uses energy; it does not produce it. Here in the 21st century, we have a shortage of energy, not technology. The shortage of energy was caused primarily by the introduction of new technologies such as the internal combustion engine. The shortage is therefore unlikely to be solved by the introduction of even more technology. More technology will simply allow us to use more energy, which will make us more dependent on technology, which will make us more dependent on energy. As the supply of energy dwindles, the technology on which we have become dependent will no longer function.

To illustrate: what do you think would happen if the average fuel efficiency of every vehicle on the road today was magically raised to 200 miles per gallon?

It doesn't take a psychic to accurately predict how we would react to this "miracle." We would continue to build our homes farther and farther away from our jobs and grow our food farther and farther away from our stores. In other words, we would increase our dependency on cheap energy. This would temporarily delay the crisis while reinforcing the underlying problem, which is a dual dependence on cheap energy and high technology.

The more dependent we are on cheap energy when the day of reckoning arrives, the more painful it is going to be, the more people are going to die, and the longer it will take us to recover from the aftermath. Consequently, increases in fuel efficiency and technology are more likely to make our situation worse, not better.

As amazing as it sounds, George W. Bush may have been correct when he said, "We need an energy policy that encourages consumption."

22. I heard some scientist has a theory that fossil fuels actually renew themselves. Is there any truth to that, and if so, will it make any difference in our situation?

No.

According to Dr. Thomas Gold, author of *The Deep Hot Biosphere*, oil is formed from a renewable "abiotic" process that occurs deep within the Earth. According to many proponents of his theory, oil is actually an unlimited resource. Some of these proponents go so far as to claim, "Peak Oil is a scam."

These folks ignore both scientific data and common sense. No remotely legitimate geologist takes this theory seriously. Nor do the oil companies, who would be more motivated than anybody to find an "unlimited" source of oil. An oil company's shareholder value is largely based on the amount of proven oil reserves it has. If a company found an unlimited reserve, their stock would shoot into the stratosphere. As explained in the next question, the oil industry is taking all the actions you would expect from an industry who knows its best days are in the past. If they thought there was any chance of locating an unlimited source of oil, they would not be downsizing and merging like there's no tomorrow.

As Dr. Colin Campbell has explained, "sometimes an oilfield will appear to slightly 'refill' itself, but this is nothing more than oil from the bottom of the well leaking in from a deeper accumulation."[85] In other words, it is not being newly created as many would hope.

Furthermore, this phenomenon has only been witnessed in the occasional well. When it does occur, the rate of "refilling" is so slow it may as well not be taking place at all. Thus, even if Gold's theory has some truth to it, it is practically irrelevant as oil fields aren't regenerating themselves anywhere near fast enough to prevent a civilization wide crisis.

23. How is the oil industry reacting to Peak Oil?

The oil industry is taking a three-pronged approach to Peak Oil: closing gas stations, merging companies, and downsizing workforces.

A. Do the oil companies plan on closing gas stations?

Yes.

The November 2003 issue of the Association for the Study of Peak Oil Newsletter reports that:

> The September 2003 issue of *World Oil* reports that Chevron-Texaco plans to dispose of 550 filling stations in the United States; 900 in Asia and Africa; retail and refining operations in Europe, South America, Australia and the Middle East; and exploration and production holdings in North America, the North Sea and Papua.[86]

Closing gas stations is a logical step if you expect there to be less gas to sell in the foreseeable future.

B. Have the oil companies been merging?

Yes.

In 1995, Petroconsultants Pty., Ltd., one of the largest and most respected oil industry analysis and consulting firms in the world, released a document called, "World Oil Supply: 1930-2050." This report was distributed to oil industry insiders and cost a whopping $32,000 per copy. It predicted oil production would soon peak and go into terminal decline. The oil industry took note. Within a few years, oil companies began merging as though they were living on borrowed time:

December 1998:	British Petroleum and Amoco merge.
April 1999:	BP-Amoco and Arco agree to merge.
December 1999:	Exxon and Mobil merge.
October 2000:	Chevron and Texaco agree to merge.
November 2000:	Russia's Lukoil announces it will buy Getty Petroleum.
November 2001:	Phillips Petroleum and Conoco agree to merge.
September 2002:	Shell acquires Pennzoil-Quaker State.
February 2003:	Devon Energy acquires Ocean Energy.
March 2003:	Frontier Oil and Holly agree to merge.
March 2004:	Marathon acquires 40 percent of Ashland Corporation.
April 2004:	Westport Resources acquires Kerr-McGee.

By July 2004, some analysts were suggesting BP-Amoco consider "the mother of all mergers" with Shell.[87]

Mergers involving smaller companies are too numerous to list. As early as 1999, Investment bank Goldman Sachs commented, "The great merger mania is nothing more than a scaling-down of a dying industry in recognition of the fact that 90 percent of global conventional oil has already been found."[88]

C. Have the oil companies been downsizing their workforces?

Yes.

In 1982, energy companies employed 1.65 million people. By 1999 the number had dropped to 641,000. On average, the largest oil companies eliminated 5.2 percent of their workforce every year between 1988 and 2000.[89]

A Labor Department study found that more than 65 percent of workers in the oil and gas industry are between ages 35 and 54, while just a "small percentage" were in their twenties.[90]

In 1998, major oil companies employed 83,000 people in the exploration and production sector. One year later, that number had dropped to 57,000.[91] Between 1997 and 1999, the oil and gas industry shed 60,000 exploration and production jobs.[92]

Universities have taken note of these trends and made the appropriate cuts in their oil-related courses of study. For instance, in 1986, 102 students graduated from the Colorado School of Mines with bachelor's degrees in petroleum engineering; in 2001, there were 34.[93] Similarly, at the University of Texas, about 180 petroleum engineering students graduated in 1982, compared with 34 in 2001.[94]

It only makes sense for an oil company to drastically cut its exploration and development workforce if it believes there is drastically less oil to explore and develop.

D. Why are the oil companies taking such drastic actions?

These are the actions of an industry who knows it's heyday is over. Much like a senior citizen who knows they will no longer be able to produce as they did during their prime, the oil industry is downsizing its operations and minimizing its activities. This way, it can have as profitable and comfortable a decline as possible. If "actions speak louder than words," the actions of the oil companies over the last 5-6 years speak volumes about the coming crisis.

24. People have been saying we'd run out of oil since as long as I could remember. In fact, didn't the Club of Rome make this exact same prediction back in the 70s?

To a certain degree, it's true: people have been warning us about "running out of oil" since we first discovered the stuff. What's different this time is the warnings are backed up by an absolutely overwhelming amount of evidence:

1. Oil discovery peaked over 40 years ago and has dwindled to almost nothing in recent years.

2. As per Richard Duncan's calculations in a May 2003 *Oil and Gas Journal* article entitled, "Three World Oil Forecasts Predict Peak Oil Production," 99 percent of the world's oil comes from 44 oil-producing nations. Twenty-four of these nations are now in permanent decline.[95] Many of them are declining at alarmingly rapid rates.[96] Saudi Arabia may even be in decline now (explained later).

3. With the exception of the Middle East, world production peaked in 1997. This includes most of the technologically advanced nations of the world: the US peaked in 1970, Russia peaked in 1987, the UK peaked in 1999.

4. Production of conventional oil has already peaked.

5. Even the most optimistic analysts are now predicting a crisis within our lifetimes.

As far as the Club of Rome: in 1972, they released a study titled *The Limits of Growth* which used sophisticated MIT computer program to predict that by 2072, modern civilization would begin a mass die off due to either resource depletion or pollution.

Often, whenever somebody makes an "end of the world"-type prediction, they are derided as a "Club of Romer." This is really unfortunate as we are right on track to fulfill their predictions. Contrary to popular misquotation, the COR never predicted "we would run out of oil by the year 2000" as so many economists and Peak Oil deniers would have you believe.

On a personal note: whenever I give a public talk about Peak Oil, there is always at least one individual who insists on proclaiming, "Matt, people like you have been claiming the 'end of oil is around the corner' for years! I didn't listen to *The Limits to Growth* folks 35 years ago and I'm not going to listen to you now! Go to hell, you Malthusian eco-freak."

Whenever this happens, I can't help but think of an obese, middle-aged, sexually promiscuous, hard-drinking smoker who insists on boasting, "Hah! The doctors have been telling me for years all this eating, drinking, smoking and fucking is going to kill me. But I'm still here! I didn't listen to the doctors back then and I'm not going to listen to you now! Go to hell, you muscle-bound health-freak."

In both instances, my usual response is to call the individual a "Girly-Man."

25. We had oil problems back in the 1970s. Will this be as bad as that?

No. It's going to be drastically worse.

The oil shortages of the 1970s were the results of political events. The coming oil shortage is the result of geologic reality. You can negotiate with politicians. You can bribe, blockade, or invade Middle East regimes. You can't do any of that to the Earth.

As far as the US oil supply was concerned, in the 1970s there were other "swing" oil producers like Venezuela who could step in to fill the supply gap when OPEC countries cut their production. Once **worldwide** oil production peaks, there won't be any swing producers to fill in the gap. The crisis will just get worse as time marches forward.

The oil shortages of the 1970s were comparable to driving a car over speed bumps. The car is simply forced to temporarily slow down while the passengers inside experience a few minor jolts. The coming oil shortages, however, will be comparable to driving a car straight into a brick wall at 90 mph. In a best-case scenario, the resulting crash will be very expensive, quite messy, and thoroughly painful for all involved. More likely however, the car will end up totaled while most, if not all, of the passengers inside will end up dead.

26. **The "end of the world" is here, once again. So what's new? Y2K was supposed to be the end of the world, and it turned out to be much ado about nothing.**

What's new is that this is the real thing. It isn't a fire drill. It isn't paranoid hysteria. It is the real deal.

Peak Oil isn't "Y2K Reloaded." Peak Oil differs from previous "end of the world" scenarios, such as Y2K, in the following ways:

1. Peak Oil is not an "if," but a "when." Furthermore, it is not a "when during the next 1,000 years," but a "when during the next 10 years."

2. Peak Oil is based on scientific fact, not subjective speculation.

3. Government and industry began preparing for Y2K a full 5-10 years before the problem was to occur. We are within 10 years of Peak Oil, and we have made no preparations for it.

4. The preparations necessary to deal with Peak Oil will require a complete overhaul of every aspect of our civilization. This is much more complex than fixing a computer bug.

5. Furthermore, oil is more fundamental to our existence than anything else, even computers. Had the Y2K predictions come true, our civilization would have been knocked back to 1965. With time, we would have recovered. When the oil crash comes, our civilization is going to get knocked back to the Stone Age. We will not recover, as there will not be enough economically available oil left to power a recovery for more than a handful of people.

27. **When you say Peak Oil may knock us back to the "Stone Age," you're exaggerating to make a point, right?**

No. When I say "Stone Age," I mean it.

Those who think we can just quietly slip back to 1765 don't understand how important energy is to our way our life. A lack of energy directly impacts our ability to acquire and make use of all resources and materials.

In a recent email, an oil engineer who reads my site explained to me that the Bronze Age was only possible because copper ores available at that time assayed at 30-50 percent metal and were therefore extractable by the primitive (low-energy) firing technologies of the day. Today, the world's best copper mines average less than 0.8 percent copper, and thus require heavy, energy-intensive, oil-powered machinery in order to extract the copper!

The same holds true for almost every resource and material known to humanity, including resources and materials such as platinum, silver, and uranium, which are necessary to up-scale alternative forms of energy such as hydrogen fuel cells, solar panels, windmills, and nuclear power plants.

We won't even be able to recycle the leftovers of industrial civilization without cheap energy, as recycling things like SUVs, computers, asphalt, etc., is extremely energy-intensive. Most recycling centers (particularly large, industrial ones) get their energy from – you guessed it – fossil fuels!

Unless you're super-rich, it's back to the caves.

28. Some expert on the news just explained the recent rise in oil and gas prices is due to terrorism in Saudi Arabia or the crisis in Russia, and concerns about "running out" are unfounded.

The "expert" conveniently failed to mention:

1. Prior to the year 2000, minor disruptions in supply due to terrorism or political unrest would have had a relatively small effect on oil prices. For instance, had terrorism in Saudi Arabia or political shenanigans in Russia interrupted the world oil supply during the mid-1990s, the oil-consuming nations of the world would simply have turned to more stable oil producers such as the UK.

    With all oil-producing nations of the world now pumping at full capacity, however, there is nobody who can significantly up their production within a reasonably short period of time. Even the Saudis now appear incapable of increasing their production by more than a small margin.

    Supply is stretched so tightly that even a minor disruption will cause a major spike in oil prices.

2. The price of oil skyrocketed from $11 per barrel in the late 1990s to over $49.80 per barrel by August 2004. That's a 450-percent increase in less than 5 years. Yet OPEC has stated over and over again the price of oil has increased only 10-15 percent due to fear of disruptions from terrorism.

Anybody who knows enough about oil to be called an "expert" is well aware of these two points. Any "expert" who fails to mention them when attempting to explain away high oil prices is lying through omission.

29. Can't the Saudis raise their production and help stave off a crisis?

Probably not, for the following reasons:

A. Is Saudi Arabia's oil production peaking?

Nobody knows for sure, but the answer is looking more and more like "yes." Matt Simmons has repeatedly raised concerns about the ability of Saudi Arabia to increase its production, stating he believes they are at capacity, in which case the world is at capacity.[97]

In recent months, a chorus of highly reputable voices in the oil and gas industry has echoed Simmons' concerns.[98] One notable voice is Dr. Ali Samsam Bakhtiari, the chief of the National Iranian Oil Company. As Dr. Bakhtiari explained in a 2003 article he penned for *Oil and Gas Journal*:

> With 100 billion barrels of crude oil produced so far, Saudi Arabia should not be far from the midway point of its proved reserves of 260 billion bbl—that means just 10 years at the going rate of roughly 3 billion per year. Bearing in mind the 'spurious revision' of 1990 that boosted proved Saudi reserves to 257 billion barrels from 170 billion barrels, the midway point could happen even sooner than that.[99]

Efforts by the Saudi oil industry to quell the fears of both industry insiders and laymen seem to have fallen short. In April 2004, for instance, CSPAN televised a conference sponsored by the Saudi Business Council in which numerous representatives of both the Saudi and US oil industries assured the audience that there is nothing to worry about in regards to the world's oil supply and that fears of a crisis are completely unfounded. I got a nervous knot in my stomach while watching the conference because as Julian Darley of the Post Carbon Institute later pointed out, "That so many high-ranking Saudi and US officials should gather in public to tell us not to worry should be quite worrisome."[100]

A month later, the knot in my stomach became considerably tighter after reading a *BBC* report on the May 2004 Peak Oil conference. According to the *BBC* report, Faith Birol, the chief economist for the International Energy Agency, explained there will be no oil crisis so long as Saudi Arabia can quickly raise its production by 30 percent. When the *BBC* reporter asked Birol if such a massive increase in production was actually possible, he responded, "You are from the press? This is not for you. This is not for the press."[101]

B. Could Saudi oil production plummet suddenly?

Yes.

The Saudi oil industry uses a technique known as "water injection" to extract oil from their larger (older) fields such as Ghawar. In laymen's terms, they pump salt water underneath the oil reservoirs, which pushes the oil up to the top, making it is easier to extract.[102]

Water injection raises a field's productivity in the short term at the expense of a much steeper decline once the field has peaked. Whereas a regular field might decline at a gradual rate of 2-3 percent per year once it passes its peak, a water-injected field will practically collapse in the years following its peak.[103]

In July 2004, Saudi Arabia's oil production dropped by an alarming 400,000 barrels per day, even though the Saudi oil industry claimed to be pumping at full capacity.[104] This is a very disturbing development, as it could be the first sign of an impending collapse of their oil production.

C. How easily could Saudi oil production be disrupted by terrorism?

Very easily.

According to former CIA officer Robert Baer in his recent book *Sleeping With the Devil*, "Taking down Saudi Arabia's oil infrastructure is like spearing fish in a barrel." As the June 2004 edition of *This is London* reported, "A coordinated assault could put the Saudis out of the oil business for two years and, according to Baer, that 'would be enough to bring the world's oil-addicted economies to their knees.'"[105]

The ease with which terrorists can disrupt Saudi oil production has become clear in the last two years. During 2003-04 there were 20 violent terrorist incidents in different parts of Saudi Arabia. The most gruesome of these attacks occurred on May 29, 2004, when terrorists attacked a building in Saudi Arabia's seaside resort Khobar and butchered foreign workers.

The attacks caused oil prices to temporarily spike into the $40-per-barrel range as investors became jittery.[106] Had the terrorists attacked the oil infrastructure, however, oil production would have dropped, prices would have absolutely zoomed, the housing bubble would have burst, and we would have had a total financial collapse on our hands.

If the financial trouble brought on by the attacks is deemed severe, a series of executive orders gives the president complete legal authority to declare martial law, intern citizens in work camps, confiscate all forms of transportation and communication, and to seize all food and water supplies. See Section VI for a more detailed explanation of the draconian measures the executive branch can implement to address a crisis that threatens "economic growth and prosperity."

30. How does all this tie in with Global Climate Change?

Unfortunately, it now appears we will have to deal with the implications of Peak Oil at the same time we finally have to pay the piper in regards to global climate change.

In February 2004, the Pentagon released a report on global climate change that was nothing short of horrifying. According to the report, the world may soon delve into atomic anarchy as nations attempt to secure food, water, and energy supplies through nuclear offensives.[107] The report concludes, "An imminent scenario of catastrophic climate change is plausible and would challenge US national security in ways that should be considered immediately."[108]

In June 2004, the CEO of Shell admitted that the threat of climate change makes him "really very worried for the planet."[109]

When both the Pentagon and the CEO of one of the world's biggest oil companies both openly admit climate change is an extraordinary threat to humanity, it's safe to say we've got real problems.

Unfortunately, the problems associated with global climate change will tend to compound the problems associated with Peak Oil, creating a constantly self-reinforcing loop of crop failure, energy shortages, and economic meltdown. Our ability to sustain the food supply will be greatly diminished as pesticides, fertilizers, and fuel become prohibitively expensive. At the same time, our ability to produce food without these petrochemical inputs will be severely undercut by unpredictable weather patterns.

31.   How does all this tie in with the water crisis?

Unfortunately, it's not just oil we're "running out of."

While the world has as much fresh water today as it did 10,000 years ago, population growth and industrialization have drastically increased demand. There are already about 1-1.5 billion people on the planet who lack access to sufficient drinking water. This number will only increase as the population and economy continue to grow.

Unfortunately, the energy crisis will serve to compound and reinforce the water crisis. One of the most frequently cited "solutions" to the fresh water crisis is desalination. The problem with desalination is that it is extremely energy-intensive. Generally, the energy used to desalinate water comes from fossil fuels.

Thus, as we slide down the down-slope of energy production, we will have less and less fresh water available to us on a per-capita basis. The availability of fresh water will further be impacted by the wars which will accompany the oil shortages.

For this reason, when people ask me if they should spend $10,000 on solar panels, I generally tell them, "Not until you have secured a supply of fresh water." You can live without electricity. You can't live without water.

Part III. Alternatives to Oil: Fuels of the Future or Cruel Hoaxes?

"A man is his own easiest dupe, for what he wishes to be true he generally believes to be true."
-Demosthenes c.383-322 BC

"For the great majority of mankind are satisfied with appearance, as though they were realities, and are often more influenced by the things that seem than by those that are."
-Niccolo Machiavelli

"A pleasant illusion is better than a harsh reality."
-Christian Nevell Bovee

"I'd put my money on the Sun and solar energy. What a source of power! I hope we don't have to wait until oil and coal run out before we tackle that."
-Thomas Edison

"All these imaginary panaceas turn to mists in a swamp at night when examined with the lenses of net energy and energy profit ratios. Nothing will replace what we are burning up quickly now. Every possible replacement has problems which have received little publicity."[110]
-Dr. Kenneth Watt

32. What about alternatives to oil? Can't we just switch to different sources of energy?

Unfortunately, the ability of alternative energies to replace oil is based more in mythology and utopian fantasy than in reality and hard science. Oil accounts for 40 percent of the current US energy supply and a comparable percentage of the world's energy supply. The US currently consumes 7.5 billion barrels of oil per year, while the world consumes 30 billion per year.

None of the alternatives to oil can supply anywhere near this much energy, let alone the amount we will need in the future as our population continues to grow and industrialize.

When examining alternatives to oil, it is of critical importance that you ask certain questions:

1. Is the alternative easily transportable like oil? Oil and oil-derived fuels such as gasoline are extremely convenient to transport. The ease with which gasoline is transported stands in stark contrast to the difficulty with which some of the proposed alternatives are transported. For instance, transporting hydrogen over long distances is virtually impossible since it is the smallest element known to man. As such, it will leak out of almost any container.

2. Is the alternative energy-dense like oil? In terms of energy-density, none of the alternatives to oil even come close to packing the wallop packed by oil.

3. Is the alternative capable of being adapted for transportation, heating, and the production of pesticides, plastics, and petrochemicals?

4. Does the alternative have an EROEI comparable to oil?

    Oil used to have an EROEI as high as 30. It only took one barrel of oil to extract 30 barrels of oil. This was such a fantastic ratio that oil was practically free energy. Some oil wells had EROEI ratios close to 100. In fact, at one point in Texas, water cost more than oil!

    Cheap (high-EROEI) energy has formed the basis upon which all of our economic, political, and social institutions and relationships have formed. Live in the suburbs and commute to work? You can only do so as long as we have cheap energy to fuel long-distance transportation. Met your spouse at a location more than a one-hour drive from your home or work? Never would have happened without cheap energy. Eat food shipped in from all around the world? Can't do it without cheap fossil fuel powered transportation networks.

None of the things we have become accustomed to in the industrialized world would have existed if the EROEI of oil had been as low as the EROEI of the alternatives we hope to replace oil with.

5. To what degree does the distribution, implementation, and use of this alternative require massive retrofitting of our industrial infrastructure? How much money, energy, and time will this retrofitting require?

6. To what degree does the distribution, implementation, and use of this alternative require people and institutions to fundamentally and radically alter the way they do business and live their lives? Will people and institutions be willing to make these changes prior to the onset of severe oil shocks, or will we wait until it's too late? How much of a competitive disadvantage will the businesses who first make these changes be at compared to their fossil fuel-consuming competitors? Will these businesses be able to survive this competitive disadvantage under increasingly poor economic conditions?

7. To what degree does the distribution, implementation, and use of this alternative require other resources which are in short supply? Do these other resources exist in quantities sufficient enough that the alternative is capable of being scaled up on a massive level? Are these resources located in highly unstable parts of the world? To what degree are the discovery, extraction, transportation, refining, and distribution of these resources dependent on cheap oil?

8. To what degree does the distribution, implementation, and use of this alternative require massive upfront investments in money and energy, both of which will be in short supply as the world begins to suffer from severe oil shocks?

9. What are the unintended consequences of the distribution, implementation, and use of this alternative?

We have an energy infrastructure which is incredibly mammoth, intricate, and volatile. It is inextricably intertwined with economic, political, and social systems equally mammoth, intricate, and volatile.

When you are dealing with systems this complex, even a minor change can set off a ripple of unintended and destabilizing effects. Attempting to make fundamental changes, like where you get energy from and how much you pay for it, can have disastrous effects, regardless of how well-intended the attempts are.

For instance, as Michael Kane pointed out in a recent FTW article entitled "Beyond Peak Oil," a crash/wartime scale program to manufacture a new generation of super fuel-efficient vehicles would likely worsen the world's water crisis.[111] How is that possible you ask? Simple, the average car pollutes 120,000 gallons of fresh water just during it construction.[112]

Finally, keep in mind the scope of the task at hand. This is something almost everybody new to oil depletion fails to consider. When discussing the world's energy supply, the numbers thrown around are so big, they cease to mean anything to the average person.

Here is an exercise that might help you understand the scope of what we are dealing with:

1. Take a look around the room you are in right now. Most everything you see was constructed using fossil fuels, transported using fossil fuels, and is powered using fossil fuels. If you're like most people, there is not one thing in your home that was constructed, transported or powered by alternative energy.

    (Note: According to the United Nations University, the computer on which I'm typing the manuscript for this book consumed 10X its weight in fossil fuels during its construction.[113] It is consuming more fossil fuels as I type.)

2. Now, as Richard Heinberg recommends in *The Party's Over: Oil, War, and the Fate of Industrial Civilizations*, take a seat in the center of the nearest big city and take a look around at all the cars, buildings, businesses, etc. Imagine all the energy that is being consumed.[114]

    Almost all of this energy is coming from fossil fuels. If you're lucky, you might find one building that makes use of solar panels or one car powered by biodiesel. Of course, even solar panels and biodiesel are manufactured in fossil fuel powered manufacturing plants.

3. Imagine all of the homes, roads, cars, computers, airplanes, airports, boats, buildings, farms, cities, etc., in the rest of industrialized world, the overwhelming majority of which were constructed, transported, and powered by fossil fuels.

The problem is not so much coming up with alternatives as it is coercing this monstrously large system to adapt to the limitations of those alternatives without collapsing in on itself and destroying everything and everyone who is plugged into it.

33. Can't we use coal to replace oil?

Like oil, coal is a fossil fuel. It accounts for 25 percent of the current US energy supply. While coal can be substituted for oil in some limited applications, it will only be able to cover a small percentage of the coming energy shortfall due to the following reasons:

1. Coal is about 50 percent to 200 percent heavier than oil per energy unit.[115] This makes it much more expensive and energy-intensive to transport than oil.

2. The machinery used to mine and transport coal usually runs on oil. As oil becomes more expensive, so will the extraction and transportation of coal.

3. Coal is an extremely dirty fuel.[116] According to Dr. David Goodstein, if coal use is expanded enough to cover the shortfall in energy supply brought on by Peak Oil, we can expect global warming effects so severe the Earth would become inhospitable to human life.[117]

4. Many Peak Oil "deniers" claim that because we have hundreds of years of coal in the ground, we can use it as a transition fuel. Technically, that's true: if demand for coal remains frozen at its current level, we have about 250 years' worth in the ground. However, if population growth is factored into the equation, we have only about 100 years' worth. If coal is used as a large-scale substitute for oil, we only have about 50 years' worth.[118]

   As with oil, the production of coal will peak long before the supply is exhausted. If we were to use coal as a large-scale substitute for oil, we'd probably hit Peak Coal inside of 25 years.

5. According to John Gever et al., in *Beyond Oil: The Threat to Fuel and Food in the Coming Decades,* coal used to have an "Energy Profit Ratio" of about 100. (EPR is measure of net-energy similar to EROEI.) Currently, coal's EPR is dropping rapidly. At its current rate of decline, coal's EPR will drop to .5 by the year 2040.[119] In other words, it will be an energy loser: it will take two units of coal to extract one unit of coal. When any resource requires more energy to extract it than it contains, it ceases to be an energy source.

34. What about substituting natural gas for oil?

Like oil and coal, natural gas is a fossil fuel. It accounts for 25 percent of the current US energy supply.[120] As a replacement for oil, it is unsuitable for the following reasons:

1. US natural gas production peaked around 1970. By the year 2000, US domestic production was at 1/3 of its peak level.[121] Many experts believe the impending "gas peak" will be as disastrous as the oil peak. It is therefore ridiculous to believe natural gas can be used as a substitute for oil.

2. While natural gas can be imported in its liquefied form, the process of liquefying and transporting it is extraordinarily expensive and very dangerous.

3. Gas is not suited for current airplane, boats, and heavy industrial equipment such as tractors, trailers, harvesters, etc.[122]

4. Natural gas cannot be used to provide for the huge array of petrochemicals oil is used to provide.[123]

35. What about using methane hydrates from the ocean floor as fuel?

Methane hydrates are deposits of ice-like crystals found on the ocean floor. They contain absolutely massive amounts of natural gas.[124] This has led many people to believe they will eventually serve as a replacement for oil. Unfortunately natural gas derived from methane hydrates is an unsuitable replacement for oil for the following reasons:

1. Although abundant, methane hydrates are difficult to accumulate in commercial quantities.

2. Recovery is extremely dangerous and considerably more expensive than the extraction of traditional oil and gas reserves.[125]

36. What about geothermal energy? Could we get our energy from things like volcanoes?

Less than 1 percent of the world's electricity production comes from geothermal sources.[126] As a replacement for oil, it is unsuitable due to the following reasons:

1. Geothermal energy must be harnessed from hot springs, volcanoes, or geysers.[127] It is therefore incapable of meeting the needs of most industrial nations.

2. Can't be adapted for cars, boats, airplanes, tanks, and other forms of transportation.

3. Can't be used to produce petrochemicals, plastics, or fertilizers.

4. As with other alternative sources of energy, we would need to build an entire new infrastructure to run on a fuel derived from geothermal energy.

37. What about hydrogen? Everybody talks about it so much; it must be good, right?

Hydrogen accounts for 0.01 percent of the US energy supply. As a replacement for oil, it is unsuitable due to the following reasons:

1. Hydrogen must be made from coal, oil, natural gas, wood, or through the electrolysis of water. Regardless of the source, it takes more energy to create hydrogen than the hydrogen actually provides. It is therefore an energy "carrier," not an energy source.[128]

2. Liquid hydrogen occupies four to eleven times the bulk of equivalent gasoline or diesel.[129] This makes it extremely difficult to store and transport.

3. A "hydrogen economy" would require massive retrofitting of existing transportation networks.

4. Hydrogen cannot be used to manufacture petrochemicals or plastics.

5. The cost of fuel cells is absolutely astronomical and has shown no sign of coming down.

6. A single hydrogen fuel cell requires 20 grams of platinum. If the cells are mass-produced, it may be possible to get the platinum

requirement down to 10 grams per cell. The world has 7.7 billion grams of proven platinum reserves. There are approximately 700 million internal combustion engines on the road.

10 grams of platinum per fuel cell x 700 million fuel cells = 7 billion grams of platinum, or practically every gram of platinum in the Earth.

Unfortunately, as a recent article in *EV World* points out, the average fuel cell lasts only 200 hours. Two hundred hours translates into just 12,000 miles, or about one year's worth of driving at 60 miles per hour.[130] This means all 700 million fuel cells (with 10 grams of platinum in each one) would have to be replaced every single year.

Thus, replacing the 700 million oil-powered vehicles on the road with fuel cell-powered vehicles, for only 1 year, would require us to mine every single ounce of platinum currently in the Earth and divert all of it for fuel cell construction only.

Doing so is absolutely impossible as platinum is astonishingly energy-intensive (expensive) to mine, is already in short supply, and is indispensable to thousands of crucial industrial processes.

Even if this wasn't the case, the fuel cell solution would last less than one year. As with oil, platinum production would peak long before the supply is exhausted.

What will we do, when less than 6 months into the "Hydrogen Economy," we hit Peak Platinum? Perhaps Michael Moore will produce a movie documenting the connection between the President's family and foreign platinum companies? At the same time, a presidential candidate will likely proclaim a plan to "reduce our dependence on foreign platinum," while insisting he will "jawbone the foreign platinum bosses," and "make sure American troops don't have to die for foreign platinum."

If the hydrogen economy was anything other than a total red herring, such issues would eventually arise as 80 percent of the world's proven platinum reserves are located in that bastion of geopolitical stability, South Africa.

7. It's possible to use solar-derived electricity to get hydrogen from water, but as physicist Dom Crea points out, a renewable, hydrogen-based economy will require the installation of 40 trillion dollars worth of photovoltaic panels.[131]

This is on top of the cost of mining every single ounce of platinum in the Earth, building the fuel cells, and constructing a hydrogen infrastructure. All of which would have to completed in the midst of massive oil shortages and severe economic dislocations.

8. Because hydrogen is the simplest element, it will leak from any container, no mater how strong and well insulated, at a rate of at least 1.7 percent per day.[132]

Hydrogen is such a poor replacement for oil that "Hydrogen Fuel Cells" should be called "Hydrogen Fool Cells." The "Hydrogen Economy" is a 30-year-old circus-act we never seem to tire of: In 1974, President Nixon proposed "Project Independence," which promised to end America's reliance on foreign oil. The project claimed that "hydrogen-fuelled vehicles" would be ready by 1990.[133]

The Earth Day crowd ate it up while politicians from both sides of the aisle milked it for all it was worth. Ohio Democrat Charles Vanick said, "Hydrogen offers us great potential as a fuel for the future." California Republican Robert Wilson stated, "We can now look forward to running our automobiles on water."[134]

The hype surrounding hydrogen sounds remarkably familiar, does it not? Unfortunately, the Hydrogen Economy was a myth in 1974, and it's still a myth in 2004 because the laws of thermodynamics haven't changed.

38. What about nuclear power?

Nuclear power accounts for 8 percent of US electricity production.[135] As a replacement for oil, it is unsuitable for the following reasons:

1. Nuclear power is more expensive than most people realize. A single reactor costs between 3 and 5 billion dollars, not counting the costs associated with the extraction of nuclear fuels, decommissioning, and safeguarding against accidents and terrorism. Nuclear power has only existed because the oil used to construct nuclear power plants has been so cheap.[136] The US currently has 100 nuclear reactors. In order to make up for the shortfall in energy supply created by oil depletion, we would need to build hundreds, if not thousands of nuclear reactors. The world would have to build even more. At three-to-five billion a pop, it's not long before we're talking about "real money."

2. Retrofitting our current oil-fueled transportation networks to run on nuclear-generated electricity would be enormously expensive and difficult.

3. Nuclear power cannot be used to produce plastics, pesticides, or petrochemicals.

4. Uranium is extracted and transported using oil powered machinery. As oil gets more expensive, so will uranium.

5. Waste is a major problem. Many folks are confident the waste issue can be solved through adherence to stringent safety protocols. Curiously, these folks rarely live in the neighborhoods where nuclear power plants are likely to be built.

Even if their faith in the ability of the US to adhere to stringent safety protocols is well-placed, they forget that we are talking about a global phenomena here. We cannot realistically expect nations such as China, Russia, and India to adhere to the same safety protocols as nations such as the United States and UK.

If the world turned to nuclear as a large-scale solution, large-scale accidents in these countries are almost inevitable. This would especially be the case during the initial stages of the oil shocks as the pressure to bring nuclear energy online would encourage safety protocols to be usurped.

6. Even if we were to overlook these problems, nuclear power is only a short-term solution. Uranium, too, is subject to a bell shaped production curve. Estimates of current known reserves vary, but seem to be between 25-40 years at best.[137] As with other resources, the supply of Uranium will "peak" long before the supply is exhausted.

7. "Breeder" reactors are a possibility, but present many technical challenges that we still haven't conquered. The challengers are overwhelming enough that even the extremely technologically advanced nation of Japan has abandoned its breeder reactor program. If breeder programs offered a realistic possibility of providing enough affordable energy to keep economic growth going, you can bet your bottom dollar they would not have been abandoned.

39. What about solar power?

Solar power currently supplies less than one-tenth of one percent of the US energy supply.[138] As a replacement for oil, it is unsuitable due to the following reasons:

1. Unlike energy derived from fossil fuels, energy derived from solar power is extremely intermittent: it varies constantly with weather and

the time of day. If a large city wants to derive a significant portion of its electricity from solar power, it must build fossil fuel-fired or nuclear-powered electricity plants to provide backup for the times when solar energy is not available.

2. Solar power has a capacity of about 20 percent. This means that if a utility wants to install 100 megawatts of solar power, they need to install 500 megawatts of solar panels.[139] This makes solar power a prohibitively expensive and pragmatically poor replacement for the cheap and abundant fossil fuel energy our economy depends on.

3. Oil provides 90 percent of the world's transportation fuel. Unfortunately, solar power is largely incapable of meeting these needs. While a handful of small, experimental, solar-powered vehicles have been built, solar power is largely unsuited for the vehicles such as large trucks which form the backbone of our commerce and food distribution networks.

   As mentioned previously, it is possible to use solar panels to get electricity from water, but a solar-hydrogen economy would require the installation of 40 trillion dollars of solar panels.

4. Solar energy is nowhere near dense as fossil fuel energy. In his recent book, *Out of Gas: The End of the Oil Age*, Dr. David Goodstein explains that in order to meet our current energy needs from solar power, we would need to cover 100,000 square miles with solar panels.[140]

   Such a project would require a mind-boggling level of investment and new infrastructure, in addition to the clearing of major technological hurdles.

5. Solar power cannot be adapted to produce pesticides, plastics, or petrochemicals.

6. Solar is susceptible to the effects of global climate change, which is projected to greatly intensify in the decades to come. Due to unpredictable weather patterns, even previously sunny locales such as Florida may not be able to count on a steady supply of solar energy.

7. The geographic areas most suited for large solar farms are typically very warm areas, such as deserts. This requires the energy collected by the panels to be converted to electricity and then transmitted over large distances to power more densely populated regions.

Unfortunately, heat makes electricity extremely difficult to transmit. The benefits of setting up solar farms in sun-drenched areas like the desert are largely offset by the additional costs of transmitting the electricity. The only way to overcome this problem is through the use of superconducting wires, which require copious quantities of silver, a precious metal already in short supply.

8. Virtually all solar panels currently on the market are made with silver paste. The world, however, is in the midst of a massive silver shortage that is likely to be greatly exacerbated in the years to come.

    Of all metals, silver is the best conductor of electricity. This has made it a crucial component of all computers, communications, and electrical equipment. As technology has spread, silver reserves have plummeted. The current shortage of silver is so severe many experts feel the price of silver will skyrocket from its August 2004 price of $6.50 per ounce to as high as $200 per ounce.[141] This will drive up the cost of solar power.

    To make matters worse, the only silver left is very difficult to extract and requires the use of heavy-duty, energy-intensive, **oil-powered machinery**. As oil becomes more expensive, so will the discovery, mining and transporting of silver, which will drive up the price of solar power even more.

    Furthermore, much of the world's silver reserves are located in highly unstable and unfriendly parts of the world such as the former Soviet Union.

    The same fundamentals are also true (albeit to a lesser degree) for copper, which is frequently used to conduct electricity.

9. Finally, as fossil fuels become increasingly scarce and expensive, we will have less energy to do everything, including obtaining replacement parts for things like solar panels. Even the most durable of solar panels, like all forms of technology, will require replacement parts and maintenance at some point in the future. Consequently, many of the solar panel systems in use today will likely be inoperable 40-50 years from now due to the collapse of oil-fueled manufacturing, transportation, maintenance, and distribution networks.

10. New developments in solar-nanotechnology appear quite promising, but even the scientist at the forefront of these developments, Dr. Richard Smalley, has admitted a few "miracles" are needed.[142] Nonetheless, many people cling to solar nanotechnology as our

savior. I find this rather ironic as when I hear the scientist at the top of the field use the term "miracles" I think of something along the lines of a Cubs-White Sox World Series. If that's what's it going to take to prevent a collapse, we're in big trouble.

40. What about water/hydro-electric power?

Water, i.e. hydro-electric power through building dams, currently supplies 2.3 percent of global energy supply.[143] As a large-scale replacement for oil, however, it is unsuitable due to the following reasons:

1. It is unsuitable transportation networks, particularly planes, boats, and heavy trucks.

2. It cannot be used to produce pesticides, plastics, or petrochemicals.

3. Our ability to tap hydropower is near capacity as we have already erected dams in most of the locales amenable to it.

41. What about wind power?

Like solar, wind power accounts for about one-tenth of one percent of the current US energy supply.[144] As a replacement for oil, it is unsuitable due to the following reasons:

1. As with solar, energy from wind is extremely intermittent, and is not portable or storable like oil and gas.

2. Wind cannot be used to produce pesticides, plastics, or petrochemicals.

3. Like solar, wind is susceptible to the effects of global climate change.

4. Wind is not appropriate for transportation needs.

Despite these limitations, wind power is one of the more promising alternatives to fossil fuels. It does provide net-energy, we do get some (albeit not much) of our current energy from it, and it is capable of being economically scaled up to a far greater degree than just about any of the other alternatives.

The fact that wind is one of our most promising alternatives is what makes our situation so disturbing. For instance, in order for wind to be used as fuel for transportation, the following steps have to be taken:

1. Build the wind farm. This step requires an enormous investment of oil and raw materials, which will become increasingly expensive as oil production drops.

2. Wait for X number of years while the original energy investment is paid back.

3. Construct an infrastructure through which the wind energy can be converted to hydrogen. This too requires an enormous investment of oil and raw materials. As explained previously, the development of a hydrogen infrastructure has its own set of physically insurmountable obstacles.

4. Deal with the energy shortages that will arise when the wind is not blowing. Nowadays, most wind farms are backed up by fossil fuel fired power plants. In a post-fossil fuel world, however, we won't have the ability to provide consistent power when the wind is not blowing.

5. Deal with enormous political and industrial resistance at each step.

6. Pray that we can repeat this process enough times before anarchy and war completely cripple our ability to do so.

42. What about plant-based fuels like methanol and ethanol?

Plant-based fuels will never be able to replace more than a fraction of the energy we currently get from oil for the following reasons:

1. Depending on who you consult, ethanol has an EROEI ranging from .7 (making it an energy loser) to 1.7. Methanol, made from wood, clocks in at 2.6, better than ethanol, but still far short of oil.

2. As explained previously, by 2050, the US will only have enough arable land to feed half of its population, not accounting for the effects of oil depletion. Unfortunately, the amount of land needed to grow enough corn to provide the quantity of ethanol we need is absolutely astronomical:

    A. According to Cornell professor David Pimentel, "it takes 11 acres to grow enough corn to fuel one automobile with ethanol for 10,000 miles, or about a year's driving."[145]

    B. According to Exxon Mobil CEO Lee Raymond, "If we tried to replace just 10 percent of projected gasoline use in the US in the year 2020 with corn-based ethanol, we would need to plant an area equivalent to Illinois, Indiana and Ohio with corn." As Raymond points out, that's about 1/6 of the land we use to grow crops.[146]

3. Current infrastructure, particularly manufacturing and large-scale transportation, is adaptable to plant-based fuels in theory only. In reality, retrofitting our industrial and transportation systems to run on plant-based fuels would be enormously expensive and comically impractical.

Finally, when evaluating claims about plant-based fuels, be aware of who is providing the data. I've read many glowing reports about the wonders of ethanol only to reach the bottom of the article to find out the ethanol-advocate owns a corn farm or is otherwise heavily invested in the industry.[147] As Dr. Walter Youngquist has pointed out, "the company which makes 60 percent of US ethanol is also one of the largest contributors of campaign money to the Congress."[148]

43. What about biodiesel?

The good news is biodiesel may be the best alternative we have. That's also the bad news.

A diesel-powered machine can be adapted to run on biodiesel with relative ease. This does not mean, however, that biodiesel can provide us with enough affordable energy to do more than slightly soften the coming collapse.

Generally biodiesel is produced from soybeans, which is problematic because 1) we are running out of arable land and 2) modern agriculture requires tremendous fossil fuel inputs. Our dwindling supply of fossil fuels will thus impact our ability to produce biodiesel.

A recent proposal involves building shallow pools in which to grow biodiesel-producing algae. Many proponents of this plan claim it can produce enough biodiesel to replace all transportation fuel in the US.

I'm extraordinarily skeptical of a process anytime its proponents make claims this outrageous. It immediately makes me wonder if they have a true appreciation for the mammoth and intricate relationship of oil to the world economy.

Unfortunately, even if we give the process and its proponents the benefit of the doubt, petrochemical civilization is still going to crash soon for the following reasons:

1. The world currently consumes about 82.5 million barrels of oil per day. The US consumes about 20 million of these, of which approximately 12.5 million are used for transportation. Even if biodiesel is scaled up "overnight" to replace all oil-based fuels the US uses for transportation, world demand for oil would still be 70 million barrels per day and growing at a rate of 2-4% per year. Within 5 years, we'd be back consuming as much oil as we are today. Meanwhile, the world's reserves would have been depleted to an even greater degree than they are currently. We'd still pass the peak and be well on our way down the down-slope inside of 10 years.

2. Of course, any plan to replace oil used in transportation is not going to be completed overnight, especially when we currently get next to zero percent of our energy from this source. The completion of such a plan would likely take generations considering the fact we've got 700 million internal combustion engines on the road, all of which are were manufactured in plants using tons of petrochemicals and are driven on roads made from asphalt, which is made from fossil fuels. We've also got millions of airplanes and boats that need to be adapted in addition to an entire food, water, and medical care system completely dependent on absolutely massive amounts of fossil fuels.

In short, the scope of the impending crisis is almost immeasurable! Even if the pie-in-the-sky claims of some biodiesel (and other alternative energy) proponents are technically viable, the world as we know it is still coming to a painful and abrupt end.

44. What about hemp?

Everybody's favorite biofuel suffers from the same limitations as other biofuels: lack of scalability, lack of arable land on which to grow enough of it, and a poor energy profit ratio.

Even if hemp production could be scaled up to produce a fraction of the energy provided by fossil fuels, we would just be trading "Peak Hemp" for Peak Oil. What do you do when hemp production peaks? Once it does, we're back in the same situation we are now.

In truth, the discussion of Peak Hemp is a moot point, as there is no way hemp production can be scaled up to provide more than a minuscule fraction of the energy provided by fossil fuels. I mention Peak Hemp merely to illustrate a point discussed further in Part IV: so long as we have an economy that requires growth, it doesn't matter what our primary energy source is, as production of all energy sources eventually peaks.

Hemp, however, has many properties that would make it an almost ideal food crop for post-petroleum agriculture. Unfortunately, the likelihood of widespread legalization of hemp farming in the US is, at this time, practically zero.

45. What about this new technology that can turn waste into oil?

Thermal Depolymerization (TD) is a process which can turn anything from tires to turkey guts, and even a human body, into oil. If it is scaled up to a huge degree in the very near future, it could slow our slide down the oil production down-slope in addition to helping us deal with our waste and land fill problems, but it is not the energy-source savior that many people are hoping it is:

1. Currently, there is only one plant in operation. The plant is currently producing 100-200 barrels of oil per day.[149] It's projected to produce 500 barrels per day. Either way, it's not very much compared to the 80 million barrels a day the world consumes. It can certainly be scaled up, but we would need at least 1,000 such plants pumping out 500 barrels per day just to get 500,000 barrels of oil, or about 1/160 of what the world needs per day.

2. TC is essentially high-tech recycling. Most of the waste input (such as plastic) is originally made from oil. As we slide down the down-slope of oil production, we will have less waste to put into the process.

3. Simple physics dictates that TD will never have a positive net-energy profile. The process requires energy to turn garbage into oil. The 2nd Law of Thermodynamics states energy cannot be created or destroyed. Thus, the energy obtained from the TD process will be less than the energy used to create the feedstock which went into the process.

None of these are reasons not to invest in the technology. I'm a big fan of it myself and have even considered approaching certain companies working on the process about possibly advertising on my site. But I'm not under any illusion that, even if scaled up to a tremendous degree, TD will be more than a proverbial "drop in the bucket."

The main problem I see is not with the TD process itself, but that so many people feel it offers us a way to maintain business as usual. Such thinking promotes further consumption, provides us with a dangerously false sense of security, and encourages us to continue thinking we don't need to make the radical overhaul of our civilization a priority because of "new technology."

46. What about free energy? Didn't Nikola Tesla invent some machine that produced free energy?

While free energy technologies such as Cold Fusion, Vacuum Energy and Zero Point Energy are extremely fascinating, the unfortunate reality is that they are unlikely to help us cope with the oil depletion for several reasons:

1. We currently get absolutely zero percent of our energy from these sources.

2. We currently have no functional prototypes.

3. We've already had our experiment with "free energy." With an EROEI of 30 to 1, oil was so efficient and cheap an energy source that it was practically free. In some locations, such as Louisiana, oil had an EROEI of 100 to 1!

4. The development of a "free energy" device would just put off the inevitable. The Earth has a carrying capacity. If we are able to substitute a significant portion of our fossil fuel usage with "free energy," the crash would just come at a later time, when we have depleted a different resource. At that point, our population will be even higher. The higher a population is, the further it has to fall when it depletes a key resource. The further it has to fall, the more momentum it picks up on the way down through war and disease. By encouraging continued population growth, so-called free energy could actually make our situation drastically worse.

   An analogy may be useful here: I live in a one-bedroom apartment. Let's pretend that tomorrow the energy fairy comes along and installs a free-energy device in my apartment. With the device running, I can use all the energy I want for free. Not only that, but it magically pays the rent and keeps the refrigerator full of food. Time for me to have all my friends move in with me?

   No, because my apartment still has only one bathroom. If 15-20 people move in with me, there's going to be shit all over the living room, free energy device running or not.

5. Even if a functional free energy prototype came into existence today, it would take at least 25-50 years to retrofit our multi-trillion-dollar infrastructure for such technology.

6. One can only wonder what damage we would do to ourselves if given access to free energy. We discovered oil, an amazingly powerful source of energy, and 150 years later we are closer to destroying ourselves than ever before. What do you think we will do to ourselves if we gain access to an even more powerful source of energy?

Another analogy may be useful here: say you give a young man access to a one-million-dollar bank account on his 18th birthday. Do you think he is going to handle it responsibly? My guess is no. If he's anything like I was at 18 (or even today), he's going to blow it all on expensive liquor, wild strippers, and fast cars.

In other words, he's going to consume and screw himself into oblivion, which is exactly what the human race has been doing to itself since discovering oil.

What do you think will happen if, upon depleting his one-million-dollar bank account, the young man gains access a bank account with one-billion-dollars in it? Most likely, he will continue consuming and screwing until he completely destroys himself and all those around him.

We will likely do the same thing if we ever gain access to an energy source even more abundant and powerful than oil.

47. What about using a variety of alternatives? If we use a little of this and a little of that, can't it add up?

Absolutely.

If we find a massive amount of political will, unprecedented bipartisan and international cooperation, gobs of investment capital, a slew of technological breakthroughs, and about 25-50 years of peace and prosperity to implement the changes, we might be able to produce the energy equivalent of 3-4 billion barrels of oil from alternative sources. That is about as much oil as the entire world consumed per year prior to World War II! But it is only about 10 percent of what we need currently, and an even smaller percentage of what we will need in the future.

48.   Are these alternatives useless then?

No, not at all. Whatever civilization emerges after the crash will likely derive a good deal of its energy from these alternatives. All of these alternatives deserve massive investment right now. The problem is no combination of them can replace oil, no matter how much we wish they could. All the optimism, ingenuity, and desire in the world can't change the fact two plus two is four.

None of the alternatives can supply us with enough energy to maintain even a modest fraction of our current consumption levels. Even in the best-case scenario, we will have to accept a drastically reduced standard of living. To survive, we will have to radically change the way we get our food, the way we get to work, what we do for work, the homes we live in, how we plan our families, and what we do for recreation.

Put simply, a transition to these alternatives will require a complete overhaul of every aspect of modern industrial society. Unfortunately, complex societies such as ours do not undertake radical changes voluntarily or preemptively. Nor do they attempt to solve their problems by simplifying or downsizing things. Instead, as Joseph Tainter explains in *The Collapse of Complex Societies*, when faced with large-scale problems, complex societies (such as ours) tend to gravitate towards increasingly complex solutions, which ultimately make the original problems much worse.

The fact that alternative energies are incapable of replacing fossil fuels seems to be an extremely tough pill to swallow for almost everybody except physicists and engineers. In my experience, everybody else insists that with enough political will, ingenuity, and elbow grease, we can somehow make the transition to alternative fuels.

I'm sorry, folks, but we can't. Without mammoth amounts of fossil fuels, there is simply no way we can run a society that even comes close to resembling what we are accustomed to for more than a handful of (super-rich) people. The physics of renewable energy are absolutely pathetic compared to the physics of fossil fuels! The numbers just don't add up, no matter how much we wish they would.

If you're thinking of sending me an email telling me to "go to hell" because you're positive that, with enough "American ingenuity," alternative energies can take the place of fossil fuels, don't bother. Filling my inbox with hate-filled vitriol is not going to change the laws of thermodynamics.

49. I know the physics and math of renewable energy don't add up, but what if some miracle alternative or combination of alternatives comes online and replaces 90 percent of the energy we lose as a result of Peak Oil? Would that prevent a total collapse?

No.

This is where things start to get scary. In order to understand why, you need to have a basic understanding of how money and energy interact. The average person concerned about Peak Oil tends to completely miss this even though it is **the key issue.**

The following explanation, **while considerably oversimplified**, should help illustrate the enormity of the problem to those of you unfamiliar with the connections between money and energy.

(Note: This question involves issues also discussed in Part IV: Issues of Economy, Technology and the Ability to Adapt.)

A. How is money created?

Money is created when banks loan it into existence. They simply make an entry in their computer and the money is "born." Banks are able to continually make more and more loans because the people who previously took out loans were able to pay back their loans plus interest.

To illustrate: let's say the bank loans me $100.00. I pay it back plus $6.00 in interest. I obtained the $106.00 by selling goods or services to people who had also taken out loans or who had obtained money from people who had also taken out loans. They used the money the bank loaned into existence to pay me $106.00 for the goods I sold to them. The bank then uses the interest I paid to them to make more loans to other people who then come back and buy more of whatever I'm selling.

B. What is money a symbol for?

Money is really just a symbol for energy. Remember how 9/10 calories you eat comes from fossil fuel energy? This means the $5.00 bill you use to pay for your 1,000-calorie hamburger is really just a symbol for the 900 calories of embodied fossil fuel energy in the hamburger.

C. What would cause our monetary system to collapse?

Our monetary system works as long as there is an excess of embodied energy constantly entering the economy. For the past 500 years, we've had a constantly expanding base of energy with which to fuel our economy.

This excess of embodied energy enabled people to pay interest on their loans. Paying interest is what keeps the system churning. If there isn't an excess of embodied energy entering the system, however, I won't be able to find enough extra embodied energy to pay the bank the $6.00 of interest I owe on my $100 loan. The bank then can't make a new loan to you. You are then unable to buy the hamburger from the local hamburger joint. The hamburger joint then goes out of business and defaults on its loan to the bank. The bank is then unable to make loans to other people. These people are then unable to buy goods and services or pay their employees. The process just keeps compounding itself until the whole system dissolves.

The whole process resembles dominoes precariously arranged in constantly enlarging, interconnecting circles. The larger the circles get, the more dangerous the system becomes.

D. Why won't miraculous developments in alternative energy prevent this system from collapsing?

To illustrate just how dangerous a system we've set up, let's pretend you want to open a small computer store in the next few years. You go to the bank and take out a $10,000 loan. The loan needs to be paid back within a year, in full, plus 10 percent interest.

In order to pay back the principle plus interest, you need to sell $11,000 worth of computers. If each computer sells for $500, this means you need to sell 22 computers over the next 12 months in order to pay back the principal plus interest.

Keep in mind, an average desktop computer consumes 10 times its weight in fossil fuels during its construction. If the computers you're selling each weigh 50 pounds, what you're really selling is 500 pounds of fossil fuel energy that has been converted into a computer.

In other words, in order to pay back the principal plus interest, you need to sell 22 computers, each of which is 500 pounds of embodied fossil fuel energy for a total of 11,000 pounds of embodied fossil fuel energy.

Now let's say the oil crash hits and fossil fuels are no longer available. Miraculously, however, alternative energy is instantly and seamlessly scaled up to replace an astonishing 90 percent of the energy we used to get from fossil fuels. This is a true miracle as even the most optimistic and enterprising individuals in the alternative energy industry never thought such a feat would have been remotely possible.

(Note: fossil fuels will not all of a sudden become unavailable. Neither will alternative energies be scaled up so quickly or to such a significant degree. I'm condensing the time frame to one year to simplify what can be a very difficult-to-understand topic.)

Realizing that folks like me had not accounted for miracles from the energy fairy when we made our "end-of-the-world" predictions, you utter those famous last words, "See, the doomsayers always turn out to be wrong."

Then the phone rings. It's the computer manufacturer. He explains he cannot fulfill your order for 22 computers. Like everybody else, he now has access to only 90 percent of the embodied energy previously available to him. Consequently, he can only build 20 computers to sell to you.

This is really bad news. Obtaining 90 percent of the embodied energy (computers) you need is not good enough! Unless you obtain 100 percent, you will not be able to pay back your loan at the end of the year. You could raise your prices, but this would push demand for your computers down and you would sell fewer. You would still end up defaulting on the loan.

To make matters worse, the computer manufacturer is going to have to charge you twice as much for each computer. It turns out copious quantities of silver were used to scale up the use of wind and solar energy. This drove the price of silver through the ceiling. Silver is an absolutely indispensable metal for all electrical devices such as computers and a mandatory catalyst for the production of all plastics. Consequently, the computer manufacturer has no choice but to raise his prices.

You go back to the bank to take out an even bigger loan in order to pay for the extra cost of the computers, but the bank denies you. Now that society has access to only 90 percent of the embodied energy previously available, the bank doesn't expect many people to be able to acquire enough money (embodied energy) to repay their loans plus interest. So they more or less have stopped making loans.

You go back to the computer manufacturer and buy 10 computers. You figure this is better than nothing.

You stock the computers in your store, but no customers come in to buy them. Since the bank has stopped making loans, none of your potential customers can obtain credit on which to buy a computer.

Unfortunately, all businesses find themselves in the same situation you're in: on average, they can only obtain 90 percent of the embodied energy they need to pay off their loans. The computer manufacturer, for instance, needed to sell you 22 computers in order to pay back his loan to the bank. Since he was only able to sell you 10 computers, he's going to default on his loan. So are all the other businesses that were dependent on his business.

The employees of all these businesses now find themselves unable to buy food. Consequently, farmers can only obtain 90 percent of the money (embodied energy) they need to pay back their loans. They default on their loans and start going out of business. The trucking companies were dependent on the farmers paying them money (embodied energy) so they could pay back their loans. Without enough money (embodied energy) from the farmers, the trucking companies start defaulting on their loans as well.

The more businesses that default on their loans, the fewer loans the banks can make to other businesses. The fewer loans the banks make, the less money there is in circulation, the harder it gets for companies to pay back their loans.

At this point, if the banks keep loaning money into existence without any corresponding increase in the embodied energy available to the economy, hyper-inflation will ensue. This is what happened in the 1920s in Germany. I don't need to remind you how that turned out.

Pretty much every business dependent on energy — which is to say all of them — is now simultaneously collapsing, even though alternative energies had been seamlessly scaled up to a miraculous degree.

Within a few months, a bankruptcy pandemic erupts. This is followed by bank runs. Then people start hoarding food and water. Unemployment skyrockets, the tax base diminishes, and government is no longer able to provide fire and police services. Marauding gangs begin forming.

Before long, the entire global financial system dissolves and civilization as we know comes to an end.

E. What the heck happened?!

As the global financial system collapses, everybody finds scapegoats to blame. What the finger-pointers don't realize is our fate was sealed hundreds of years ago when the modern interest-bearing monetary system first developed! There's nobody to blame but banks from the 1500s!

It's simple: once interest is charged, a perpetual growth machine is created. This perpetual growth machine requires an energy supply that is constantly expanding. Once the energy supply stops expanding, the machine implodes.

It's as though we've been worshipping a god who demands we bring him a constantly increasing supply of animal sacrifices. The moment we provide him one animal sacrifice fewer than we provided him previously, he will forsake us. When that day comes, we will cry out, "But, almighty god of capitalism, we're working our hardest to bring you what you ask for. Is 90 percent not good enough?" Unfortunately, the god of capitalism does not hear the lamentations of those who fail to pay interest.

Maybe this is why both the Bible and the Qur'an repeatedly and forcefully admonish their readers to never use an interest-bearing monetary system. Perhaps higher consciousness attempted to warn us of the dangers of fractional reserve banking via messages in our two most popular spiritual texts. It seems we've been too busy unquestionably worshipping the golden calf of market economics to pay attention to reality.

F. Why aren't the experts and governments addressing this issue?

As you can see, dealing with the oil crisis requires much more than just finding a replacement for oil. It requires replacing a growth-based monetary system with a steady-state system. Few people in the modern world have any experience implementing or dealing with such a system. None of the so-called "experts" you see on television or read in the papers have any idea how to address this. Naturally, they can't bring themselves to admit they have no idea how to handle this problem, so they simply deny its existence.

I firmly believe the collapse of modern economics will precede the collapse of the oil supply. Once the banks realize energy production is peaking, it's "game over," because everybody in the banking world knows without excess energy the whole system collapses.

One of the reasons governments cannot bring themselves to plan or prepare for Peak Oil is because they understand the enormity of the problem. On the other hand, people who insist governments address Peak Oil by launching super-sized alternative energy programs are, most often, clueless.

50.   Fine, but what if space aliens or angels come down and give us an alternative source of energy that easily replaces oil and can supply a constantly increasing amount of energy? Wouldn't that prevent a collapse?

No.

The US dollar is the reserve currency for all oil transactions in the world, hence the term "petrodollar." In short, this means that whenever anybody buys oil, anywhere in the world, they have to pay with dollars. Thus, the wealth from all oil transactions cycles into the US economy. The strength of the US economy is now entirely dependant on the strength of the petrodollar as the US manufacturing and industrial base has been dismantled and shipped to China, India, Mexico, and the Philippines. The petrodollar is one of the few things we have left with which to support our economy.

If such an alternative source of energy came online, oil purchases would drop, the petrodollar would collapse, and the US would descend into economic anarchy. The US would react (probably preemptively) to the widespread implementation of this alternative by plunging the world into a series of currency-wars unlike anything we have ever imagined.

If you wondered why the Bush administration was so amazingly determined to go to war, now you know: according to the map they are reading from, we're going to be fighting oil wars, currency wars, or both.

Unfortunately, the US is truly wedded to oil, with little possibility of an annulment or divorce. As they say, "till death do us part."

Part IV. Issues of Economy, Technology, and the Ability to Adapt.

"Our present economic system is... little more than a well-organized method for converting natural resources into garbage."
-Jay Hanson

"Higher oil prices affect the economy, but it's not like a heart attack. It's more like a cancer that just metastasizes across the whole spectrum."
-Phillip Verlager, Institute for International Economics

"Facts do not cease to be facts simply because they are ignored."
-Aldous Huxley

"Hummers are not the problem, and Hybrids are not the solution."
-Matt Savinar

"Thinking you can turn the Titanic around after it's already hit the iceberg and split in two is a tad naïve."
-Matt Savinar

51. **I don't think there is really anything to worry about. According to classical economics, when one resource becomes scarce, people get motivated to invest in a replacement resource. When the price of oil gets too high, renewable energy will become profitable and companies will begin investing in it.**

Classical economic theory works great for goods within an economy. Relying on it to address a severe and prolonged energy shortage, however, is going to prove disastrous. There are several reasons why:

A. *Is classical economics fundamentally flawed?*

Yes.

As Jay Hanson has pointed out in his article, "Five Fundamental Errors," classical economics has several fundamental flaws that prevent it from being able to appropriately react to severe natural resource shortages.[150] The most egregious, perhaps foolish, of these flaws is that the typical economist believes the economy is in ultimate control of the environment, not the other way around. To be sure, the economy is capable of exerting an effect on the environment, but as the saying goes, "Mother Nature bats last." The economy might demand the environment supply it with a resource, but ultimately the environment has the ability to restrict the economy's access to that resource. Generally, this is accomplished through positive feedback mechanisms such as nature-wide bell curve shaped extraction rates: the more of a particular resource the economy takes from the environment, the more the environment restricts the economy's access to that resource. Economists believe that with enough money we will find a way around these mechanisms.

As Hanson notes, a notable example of this uber-faith in the power of the money comes from Morris Adelman, whose ideas have had tremendous influence in the oil and gas industries.[151] On page 483 of his book, *The Economics of Petroleum Supply*, Adelman writes, "There are plenty of fossil fuels and no limit to potential electrical capacity. It is all a matter of money."

On a similar note, economist Julian Simon has explained that due to advancements in technology, the human population can go on increasing forever.[152]

The idea that "there are no limits" is downright juvenile. As physicist Stephen Hawking has pointed out, if population growth continues at its current rate, by the year 2600 the entire planet will be covered by human beings standing shoulder to shoulder and electricity use will make the Earth glow red-hot.[153] Physicists like Hawking understand the Earth is a **finite sphere with finite resources**, but economists like Simon seem to insist otherwise.

Perhaps the following analogy will help illustrate what happens in the "real world" when a population confined to a finite environment depletes its key, life-giving resource:

A young economist and a bearded, homeless guy are placed in a sealed "rat hole" out in the middle of the desert. The economist brings with him his trusty copy of *The Wealth of Nations* by Adam Smith, an autographed picture of John Maynard Keynes, and his checkbook. The homeless guy, having once been very wealthy, brings with him his trusty 9-millimeter handgun, a picture of him and his former best-friend Don, and $750,000 in cash.

The only thing the two men have to eat down in the rat hole is a loaf of bread.

Since 1) there's no limit to bread production, 2) it's all about money, and 3) the bearded man has lots of cash and the economist has lots of checks, finding a replacement resource for the rapidly dwindling supply of bread shouldn't be a problem, right?

It doesn't take a genius to figure out how this story is going to end. Our young economist can pray all he wants to the Holy Trinity of Smith, Keynes, and Greenspan, and run his fingers over his checkbook as though it were a string of prayer beads, but it's not going to do him much good as the supply of bread dwindles. Down in the finite and hostile environment of the rat hole, the only part of his economics book that's accurate is the part about "demand destruction."

B. Do any scalable substitutes for oil exist?

No.

As explained in Part III, none of the current alternatives to oil, or combination thereof, can deliver anywhere near the amount of net energy delivered by oil. It's theoretically possible that a substitute or combination of substitutes for oil will be found, but implementation of those (yet unknown) substitute energy sources will present Herculean challenges and may instigate a collapse in and of itself.

C. Will market indicators come in time for us to attempt to make changes?

No.

Once we pass the peak, oil production will decline by **at least** 1.5-3 percent per year. At the same time, demand will continue to increase by 1.5-3 percent per year as the world's population continues to grow and debt continues to require servicing. This means that in the first year after we pass the peak, the world will experience a 3- to 6-percent shortfall in oil supply.

In order to understand just how cataclysmic a 3- to 6-percent yearly shortfall is, consider the following:

1. The oil shocks of the late 1970s resulted from shortfalls in oil supply in the 4-5 percent per year range.

2. Within 15 years of the peak, depending on the rate of decline, the shortfall in oil supply will have compounded itself to a market-shattering 30-60 percent or more. The shortfall will keep compounding itself as time goes on.

Even if we had scalable alternatives to oil, it would take a minimum of 25-50 years to retrofit our industrial infrastructure, manufacturing base, and transportation/food distribution networks to run on these fuels. You can't just turn the Titanic around after it's already hit the iceberg! As John Gever explains in his book *Beyond Oil: The Threat to Food and Fuel in the Coming Decades*, if we wait for the market to react to the oil shocks before aggressively pursuing alternatives, industrial civilization will experience a 75% loss in energy availability to non-energy sectors of the economy.[154]

In effect, all of our energy would have to be channeled towards developing alternative sources of energy. The development of alternatives would come at the expense of other endeavors such as food/water distribution, national defense, police/fire services and health care.

Even if we were willing to suffer through a 75% loss in the availability of energy in order to scale up the alternatives, there is no guarantee we will be successful, especially given how most people are going to react to the energy famine. When push comes to shove, do you really think most Americans are going to get together to build solar panels and live happily ever after? I suspect most will reach for the nearest shotgun, not solar panel.

We are going to wake up one day and really regret that we waited for the market to solve this for us because once the price of oil gets high enough that people begin to seriously consider alternatives, those alternatives will become too expensive to implement on a wide scale. Reason: oil is required to develop, manufacture, transport and implement oil alternatives such as solar panels, biomass, and windmills.

There are many examples in history where a resource shortage prompted the development of alternative resources. Oil, however, is not just any resource. In our current world, it is the precondition for all other resources, including alternative ones. To illustrate: as of the summer of 2004, a barrel of oil costs about $45, which is 80 percent more than the "ideal" price of about $25 per barrel. It would cost in the range of $100-$250 to get the amount of energy in that barrel of oil from renewable sources.[155] This means an energy company won't be motivated to aggressively pursue renewable energy until the cost of oil doubles, triples, or quadruples from its already dangerously high price. At that point it will be too late: our economy will be completely devastated. Our ability to implement whatever alternatives we can think of will be permanently eliminated.

In pragmatic terms, this means if you want your home powered by solar panels or windmills, you had better do it soon. If you don't have these alternatives in place when the lights go out, they're going to stay out.

D. Conclusion

The "invisible hand of the market" is about to bitch-slap us back to the Stone Age.

52. What about the whale oil crisis of the 19th century? The market solved that crisis. What makes you think it won't solve this one?

During the early 1800s, people used whale oil to light their lamps. As the whale population shrank, a crisis emerged in the early 1830s. Between 1831 and 1854, the price of whale oil rose 540 percent. As the price of oil rose, people began to conserve. The high price created incentives for investors to come up with alternative sources of lamp fuel. As a result, kerosene was invented, and the whale oil crisis was ended.

People who insist on comparing the oil crisis of the 21st century with the whale oil crisis of the 19th century are ignoring the following facts:

1. In the 19th century, people didn't use whale oil for much of anything other than lighting their lamps. They didn't use it for transportation, to power their food supply, to pump fresh water, to produce consumer products or to power the military. Whale oil was nowhere near as important to the civilization of the 19th century as oil is to the civilization of the 21st. If the price of oil rose 540 percent, do you

think there is any way we could avert a crisis of unimaginable proportions?

2. Whale oil was not a prerequisite for an alternative source of power. People did not need massive quantities of whale oil to create kerosene. The increase in the cost of whale oil did not cripple the ability to produce alternatives. Today, we do need massive amounts of oil to come up with alternatives to oil. As oil prices rise, our ability to implement alternatives will be crippled.

For these reasons, using the whale oil crisis of the 19th century as proof the market will solve the oil crisis of the 21st century is silly at best, and disinformation at worst.

53.   The oil companies will come up with a solution to keep making money, right?

The oil companies don't need to come up with something to replace oil as they are likely to profit handsomely from the crash so long as they make the appropriate cuts in their staffs and exploration budgets. How? Simple. Say, for example, that in February 2004, it takes $10 to extract and refine a barrel of oil. If a company sells that same barrel in March 2004, they will likely fetch about $38 for it. However, if they wait until the oil crash hits hard, they may be able to sell that same barrel for considerably more.

It probably comes as no surprise to you that many energy companies were reporting record profits throughout the spring and summer of 2004, just as oil and gas prices were skyrocketing. One journalist commented, "The oil companies aren't just making money, they're printing it!"

To be perfectly blunt, expecting the oil companies to save you from the oil crash is about as wise as expecting the tobacco companies to save you from lung cancer. Corporate officers are **bound by law** to do what is in the best interests of the corporation, so long as their actions are legal. Their legal obligation is to make money for the company. It is not to save the world, not to serve their country, not to clean up the environment and most certainly not to make sure you and I continue living the comfortable existence we have grown accustomed to.

The fiduciary responsibility of the corporate officers is to the company above all else. For all intents and purposes, this means it is practically illegal for an oil executive to divert a large portion of the company's resources to renewable energy because such a diversion would severely hurt the company's bottom line. The CEO would likely find himself on the wrong end of a lawsuit by the company's shareholders, and possibly even in jail.

Occasionally, a company will stroll out a "renewable energy" initiative, or run a few commercials about their alternative fuel programs. Typically, however, the company's foray into alternative energy accounts for a miniscule fraction of their overall business. While these efforts are certainly better than nothing, they are in effect more "sizzle than steak."

Expecting the oil companies, the government, or anybody else to solve this problem for us is simply suicidal. You, me, and every other "regular person" needs to be actively engaged in addressing this issue if there is to be any hope for humanity.

54. Is it possible the oil companies are sitting on some technology that they're going to bring to the market? Or maybe the government is hiding some technology that can replace oil?

No.

Any company in possession of such technology would see their stock soar on Wall Street as soon as they announced it. They would destroy their competition overnight. There would be no reason for them to keep it under wraps. If the companies were in possession of such technologies, they wouldn't be downsizing and merging at the rate they are.

Similarly, any President who announces that we are in possession of a technology that could eliminate our reliance on foreign oil would have a holiday named after him. Additionally, the technology would likely have military applications, which the President could use to intimidate the rest of the world. Again, there would be no reason to keep such a development under wraps.

Even if such technology did exist, it is doubtful it could do us much good for all the reasons explained in Part III.

Furthermore, there are biological and genetic factors at work here that have nothing to do with the oil companies. Even if the oil companies were abolished in favor of hemp companies, we would still be in the same situation.

The reindeer on St. Matthew Island, for instance, experienced a population crash even though they had no such thing as a corporation. Once any species overshoots its resource base, it will collapse, regardless of whether it has big corporations to blame or not.

The population crash we are about to experience is as much the product of our biological wiring as it is the product of greedy oil companies.

55. I think you are underestimating the human spirit. Humanity always adapts to challenges. We will just adapt to this, too.

Absolutely, we will adapt.

Part of that adaptation process will include most of us dying if we don't take massive action right now to decentralize and minimize our economic patterns. Adaptation for millions does not equal survival for billions.

Unfortunately, there is no law that says when humanity adapts to a resource shortage, everybody gets to survive. Think of any mass tragedy connected to resources such as oil, land, food, labor (slaves), buffalo, etc. The societies usually survive, but in a drastically different and often unrecognizable form.

Easter Island is an excellent example of what happens to a human population when a resource their civilization is built around is depleted. The islanders had one of the most socially complex and technologically advanced civilizations for their time and resource base. They were certainly endowed with as much intelligence and ingenuity as any other group of people. Yet they were unable to adapt to a critical shortage of timber until their population was reduced by 98 percent.

The same was true for Native American populations who relied on the buffalo. As European settlers killed off the buffalo, many Native American tribes no longer had access to the resource upon which they based their entire civilizations. As a result, their populations crashed.

Contrary to popular stereotypes, many of these tribes had amazingly advanced societies. They certainly possessed as much native intelligence as those of us living in the modern, industrialized world.

How ironic that a few hundred years later, the descendants of the European settlers are desperately hunting for oilfields in much the same the Native Americans used to hunt for buffalo.

56.   We'll think of something. We always do. Necessity is the mother of invention.

Yes, and lots of cheap oil has been the father of invention for 150 years. No invention has ever been mass-produced and no resource has ever been extracted or distributed globally without an abundance of cheap oil. Dealing with an energy crisis of this scope is not as simple as just "thinking of something." We are talking about the collapse of a highly complex society. Complex societies such as ours are like the Titanic: in order to change course, they have to initiate the course change a long time before any icebergs are actually visible.

The post-oil collapse of North Korea, for example, while not completely analogous to countries such as the US, shows us what happens when a complex, industrialized, nation runs into a massive oil shortage:

North Korea has never had any real oil resources of its own. During the Cold War, it imported its oil from the Soviet Union, China and Iran. When the Soviet Union collapsed in 1990, North Korea's oil supply plummeted. China, Iran and other countries were either unwilling or unable to make up the shortfall created by the Soviet collapse. The oil shortage quickly sent shockwaves through every sector of the North Korean economy, particularly agriculture. Food production quickly plummeted. The collapse of the agricultural sector was compounded by the collapse of the government and industrial sectors.[156]

By 1997, the situation was stark. That year, US Congressman Tony Hall (D-Ohio) visited North Korea and was stunned at the condition of the country. According to Hall, "Everyone is systematically starving together." Hall added that he saw "evidence of a slow starvation on a massive scale," including families eating grass, weeds and bark; orphans whose growth has been stunted by hunger and diarrhea; people going bald for lack of nutrients; and hospitals running short of food.[157]

The people of North Korea did everything they could to adapt to the oil shortages — they walked more, ate less, but nothing made that much of a difference. They had plenty of necessity, in addition to as much native intelligence and work ethic as any other people, but they were unable to come up with any inventions that even moderately alleviated their situation. Necessity is the mother of invention, but she needs some food (oil) to give birth to anything.

The entire world now finds itself in a situation similar to the one North Korea found itself in 1990. With worldwide oil shortages on the horizon, there is no one we can appeal to for more oil. The post-oil collapse of North Korea should serve as a warning to anyone who dismisses the issue of oil depletion with a cavalier, "I'm not worried — we'll think of something."

Keep in mind that we often don't find solutions to serious problems. Or we find them only after many people have died. For instance, despite all of our technology, money, and ingenuity, we have no cure for AIDS, for cancer, or even the common cold. There is no guarantee that we will come up with a miraculous solution for oil depletion.

If you were diagnosed with a life-threatening disease, would you take it upon yourself to prepare, or would you dismiss the diagnosis with, "Oh, somebody will find a cure in the next couple of years before my condition gets really bad." You need to take the issue of oil depletion just as **personally and seriously** as you would a diagnosis of a terminal disease if you are to have any chance at survival.

There is no new technology or source of energy coming to our rescue. At this late stage in the game, I seriously doubt there is even anything coming to soften the crash. Some folks reading this may be tempted to perform hours of research on the internet studying exotic techno-messiahs. While some of these technologies are quite fascinating and will no doubt receive a good deal of attention from venture capitalists looking to cash in on the energy famine, they are not going to put food on your table, get you a job, or keep the lights and heat on in your residence. For you or me to hold on to the hope that one, or a combination, of these alternatives will even soften the crisis for anybody but the super-rich is simply delusional.

57. People survived for thousands of years before oil. There is no reason we can't survive without oil.

Absolutely, as long as the population contracts to what it was before the oil age.

58. What if everybody went out and got a hybrid car? Would that help the situation?

No.

Remember, oil is used to construct cars. If everybody went out and replaced their SUVs with hybrids, the demand for oil would go up, and we would quickly exhaust what little cheap oil we have left.

If current gas guzzlers could be easily and inexpensively retrofitted for hybrid fuel systems, the situation might be ameliorated slightly. Unfortunately, even that might not help much as gains in energy efficiency are almost always offset by increases in consumption.

Furthermore, because the consumption of oil grows exponentially, even if we cut our oil consumption by 2/3, we would only be postponing the peak by about 30 years. During those 30 years, we would further deplete the world's remaining reserves.

The "hybrid solution" appeals to people because it is superficial. It doesn't require any fundamental changes in lifestyle. A problem created in large part by the automobile is not going to be solved by making a few changes to the automobile. The bicycle is the transportation technology of the future, not the hybrid vehicle.

59. You're underestimating us. Look at the odds we overcame in World War II. We figured out how to defeat Hitler; we can figure out a way to deal with Peak Oil.

We were able to defeat Hitler because we had cheap oil and he didn't. During the 1940s the US was the world's number-one oil producer and the number-one oil exporter. The US provided 6 billion of the 7 billion barrels of oil used by the Allies to defeat Germany and Japan. Cheap American oil powered every aspect of the war effort, including the research and development of the atomic bomb.

In contrast, both Germany and Japan ran out of oil. Germany's fuel crisis was so severe many of its infantry units had to abandon their tanks and mechanized troop carriers in favor of horses and bicycles. Japan resorted to kamikaze tactics, in part, because there was not enough fuel for the pilots to make return trips.

Despite massive energy shortages, both countries had developed absolutely shockingly advanced weapons. The Nazis had prototypes for all the technology you see in today's "shock and awe" campaigns: supersonic jet fighters, guided missiles, and even a prototype for a flying-wing stealth bomber! Hitler insisted, even during the last days of the war, that these "wonder-weapons" would deliver the Third Reich from defeat.

In addition to having extremely advanced technology, the Nazis and Japanese were as motivated to defeat us as we were to defeat them.

The simple fact is this: we won because we had energy and they didn't. (Thank God that was the case.)

What you must understand is this: technology requires energy; it does not produce energy. Without energy, you can't develop and deploy technology, no matter how ingenious or advanced it is. It's like trying to solve an advanced calculus problem when you're starving to death. Your chances of solving that problem without first getting enough food are about the same as industrial civilization's chance of developing new technologies without first getting some energy. You can't access your higher intellectual and creativity processes unless you have enough energy to do so!

For a bit of historical perspective, remember that Hitler got hold of the German government through questionable methods. Germany then invaded Poland under false pretenses. The invasion was the first move towards grabbing the oil-rich land in the Caucuses. The invasion involved the use of, what was then, advanced military technology. It was sold to the German people as being necessary to secure the "homeland" against communist "terrorists" who had burned down the Reichstag (German Parliament). The invasion was described as a "Blitzkrieg" which basically means shockingly fast. The German leadership was convinced the war would be a cakewalk because the German military possessed highly advanced technology while their opponents had to rely on primitive weapons and guerrilla tactics.

Within a few years, Germany ran out of oil, its energy-intensive, oil-powered military was inept, its economy was shattered, its infrastructure ruined, and its people starving and humiliated. The technology that was supposed to save them never manifested because they ran out of energy before they could develop, distribute, and deploy it.

Sound familiar?

60. If we stopped spending so much money bombing other countries and put that money towards peaceful pursuits like building schools and hospitals, wouldn't that help the situation?

Not really.

Without fundamentally changing the monetary system, it would only delay the inevitable. Don't get me wrong, the world would be a much better place with more food and fewer nuclear bombs, but replacing bombs with food is akin to rearranging deck chairs on the Titanic. We'd simply be replacing perpetual economic growth based on bombs with perpetual economic growth based on food. The problem is that no system can continue to grow indefinitely, regardless of how benevolent it is. For instance, in biology, when a healthy cell continues to grow indefinitely it gets a new name: cancer. When that cancer gets to a certain size, you either cut it out or you die.

61. So how do our leaders plan on overhauling the international banking system?

They don't, as doing so would require massive decentralization of the financial system, which would result in much less control over the world's population. In their eyes, this is not an option. Now that oil production is set to permanently decline, the only way to maintain a highly centralized financial system is with a drastically reduced population.

In other words, they plan on killing us or letting us die. They already have a euphemism for the process: "demand destruction."

Don't expect whoever is elected president to have the balls (or ovaries as the case may someday be) to propose any fundamental changes to the debt-based banking system. The last two presidents to do that were Lincoln and Kennedy. You know what happened to them.

62. Can't we conserve energy in order to buy ourselves some extra time?

Not without instituting a complete financial meltdown.

The reason is simple: we have an economy mired in debt: corporate debt, government debt, and consumer debt are all at record levels. In order to finance debt, you need economic growth. Economic growth requires a constantly increasing consumption of consumer goods, most of which are made from plastic, which comes from petroleum (oil) and are delivered by trucks, which consume diesel fuel (oil).

A truly successful conservation program would require us to drastically cut our consumption of consumer goods, which would halt economic growth dead in its tracks. This would cause indebted corporations, governments, and individuals to all slide towards bankruptcy. Banks would call in outstanding debts, businesses would close, government services would cease, and people would lose their jobs. The Great Depression would begin to look like the "good old days."

Part V. Peak Oil and US Political/Social Issues

"The only difference between Bush and Gore is the velocity with which their knees hit the floor when corporations knock at the door."
-Ralph Nader, 2000 election.

"George W. Bush is not the problem and John F. Kerry is not the solution."
-Matt Savinar

"Newspapers are unable, seemingly, to discriminate between a bicycle accident and the collapse of civilization."
-George Bernard Shaw

"You can say anything you want in a debate, and 80 million people hear it. If reporters then document that a candidate spoke untruthfully, so what? Maybe 200 people read it, or 2000 or 20,000."
-George Bush's press secretary to reporters following the 1980 vice-presidential debate

"Americans seem to think that history has come to a dead stop — with them on top of the world."
-Bill Bonner

"America is a nation without a distinct criminal class, with the possible exception of Congress."
-Mark Twain

"This election's like a beauty pageant and those are two ugly bitches."
-Unknown

63. John Kerry has a renewable energy plan. If he's elected President, will that help the situation?

No.

Kerry has stated that, if elected president, he will put $20 billion towards the development of renewable energy. The money would be used to scale up alternatives to oil such as natural gas, coal, nuclear and plant based fuels derived from corn, soybeans and other crops. Kerry's stated goal is to have 20 percent of the fuels powering US vehicles come alternative sources.[158]

The specifics of the plan were probably decided upon by Kerry's advisors as a way of tapping the attention span deficit of the average voter and the sound-bite driven nature of the modern media, not by a realistic assessment of our situation. After all, the average voter probably can't remember anything much more complex than, "20 percent in 2020."

US fuel consumption is projected to increase by as much as 50 percent in the next 15 years. By that time, the US will be importing over 75 percent of its oil. Consequently, even if Kerry's plan was technically, thermodynamically, and economically viable, we will still be consuming more foreign oil in 2020 than we are currently.

By 2020, we will be 50-75% down the down-slope of the oil production curve and will have completely fallen off the natural gas cliff. Meanwhile, there will be 8 billion people on the planet, all frantically clamoring for food grown with intensive petrochemical inputs. Major disruptions in supply due to war and weather will likely be an everyday occurrence as whatever little oil we'll have left will be coming from highly unstable parts of the globe.

In short, when you take a step back and look at Kerry's plan in the context of the immerging world crisis, the scale of the needed mobilization dwarfs Kerry's proposals by an almost unimaginable scale. In this regard, it has a lot in common with Nixon's "Project Independence," Carter's extremely ambitious renewable energy plans, and Clinton's "Million Solar Home" initiative. You know how successful those programs turned out.

None of these energy sources mentioned by Kerry are even remotely viable alternatives to oil, either individually or in combination with other sources. See "Part III. Alternatives to Oil: Fuels of the Future or Cruel Hoaxes?" for a detailed explanation of why this is the case.

Kerry is no doubt aware of the true ramifications of Peak Oil. As a high-ranking member of the Senate, he has likely had access to reports and briefings similar to the CIA report about Peak Oil discussed in the following question.

He really has no choice but to play the "get off foreign oil" charade. He certainly can't announce the truth, which is that our very way of life can not be maintained under even the best of circumstances and that we need to prepare for a massive downscaling of our lifestyles.

That would upset both the voters and the petrochemical industries that pay for all presidential campaigns.

As the *Boston Globe* reported in August 2004, Kerry practically admitted he has no intention of reducing our oil dependence when he said to a crowd in Missouri, "I want Americans to drive. You want to drive a great big SUV? Terrific. That's America."[159]

Kerry's real plan to deal with the oil crisis is simply an extension of George W. Bush's plan. He has promised to "fight the war on terror better than George W. Bush." As you will see in later questions, the war on terror is really a code phrase for the war for the world's last few reservoirs of oil.

This may come as a shock to many of you, but George W. Bush is the only President, other than Jimmy Carter, who has actually been honest with the American people about our energy situation. In May 2001, he stated, "What people need to hear loud and clear is that we're running out of energy in America. We can do a better job in conservation, but we darn sure have to do a better job of finding more supply."[160]

Say what you will about Bush, he didn't sugarcoat the problem or his plan to deal with it: he acknowledged we're running out and said he'd go get some more.

That's exactly what he attempted to do. And it's exactly what Kerry plans on doing also. Only "better than George Bush," he says.

64. How long has the US government based its foreign policy on Peak Oil?

For almost 30 years.

In March 1977, the CIA issued an intelligence memorandum titled, "The Impending Soviet Oil Crisis," which predicted that the Soviet Union's oil production would peak in 1987. The document was classified as secret until its public release in January 2001 in response to a Freedom of Information Act request. Richard Heinberg, author of *The Party's Over* and *Plan Powerdown*, has completed a thorough analysis of the document and posted his findings online at http://www.museletter.com/archive/cia-oil.html.

The methodology used by the CIA in the study is discussed in testimony given before the Congressional Budget Office. A full transcript of the discussion can be downloaded from the CBO website.[161]

The document and subsequent discussion proves our government has had a clear and detailed understanding of Peak Oil since at least the Carter administration.

It's safe to assume that understanding permeated into subsequent administrations. This is especially the case with the Bush/Cheney administration for the following reasons:

1. George W. Bush's father was vice-president during the Reagan administration. It's highly unlikely Bush I would not impart knowledge of such an important issue to Bush II. Both father and son are oil-men, and thus likely have an understanding of how oil fields are depleted.

2. George W. Bush's statement in May 2001 that "America is running out of energy" indicates he has an understanding of the situation.[162]

3. Matt Simmons, who has spoken and written extensively about Peak Oil, served as an advisor to Bush and Cheney.

People assume that because they're hearing about Peak Oil for the first time, it just recently became an issue. **What they fail to realize is our government has based much of its foreign policy on Peak Oil for almost three decades.** This means it's 30 years too late for you to write your Senator or lobby your Congressman about how to handle Peak Oil. The plan now unfolding was put in place a long, long, time ago.

65. Why haven't I heard about this on the nightly news?

Peak Oil has been reported rather extensively on the Internet. It has been getting increasing coverage in the mainstream media, but the coverage is usually confined to the back page of a newspaper, an obscure part of a news agency's Website, or a puff piece that acknowledges the high price of oil but ends by reassuring the viewer/reader, "there are alternatives," and "the real problems won't start for 15-30 years."

This was the case with the June 2004 issue of *National Geographic*. Much to my delight, the cover story was "The End of Cheap Oil." Much to my dismay, the article's overall message was "there won't be real problems for 20 years." The typical reader probably came away from the article with a feeling of "those folks up in Washington D.C. should do something about this soon," or, "Gee, some scientists should really look into alternatives," as opposed to "I need to talk to my family right this moment about how we're going to deal with this because nobody is coming to save us."

If you do hear or read about Peak Oil in the mainstream press, it will probably be from a commentator who acknowledges the crisis but ends up reassuring you business will remain as usual in the long run. The interview or article will follow a predictable path:

1. They will start off by acknowledging we have a serious problem and that a good deal of economic pain is ahead. Their tone might be a bit scary or anxiety-provoking at this point.

2. Somewhere between halfway and three-quarters of the way through the interview or news article, their tone will switch to one of cautious optimism and measured reassurance.

3. They will explain that if our leaders get serious about certain technologies, things will work out okay in the long run. They will acknowledge that these technologies have obstacles, but assert that they can be solved if we throw enough money at them.

4. They will make sure to reassure you the real problems won't start until 2015 or later and that recent oil spikes are primarily the result of terrorism or political scandal.

5. They might suggest Americans consider energy-saving techniques such as carpooling, buying a hybrid vehicle, or wearing a sweater in the winter.

6. Under no circumstance will they suggest you make any major changes in your life such as giving up your car altogether, drastically slashing your consumer spending, selling your overpriced home while you still can, or pulling your money out of the stock market before it's too late.

7. If they do give time to somebody such as Dr. Colin Campbell or Matthew Simmons, they will make sure to offset their dire warnings with comforting assertions from an "expert" who insists certain technologies or alternative energies can allow for a relatively smooth transition to a post-oil world.

There are a several reasons why you won't see brutally honest Peak Oil authors and commentators like Colin Campbell, Richard Heinberg, Dale Allen Pfeiffer or even myself in the mainstream news:

1.   Every major media corporation trades on Wall Street and is heavily invested in, or sponsored by, petrochemical interests such as the energy, transportation, pharmaceutical, and agribusiness industries.

     Who's going to be suckered by the advertisement for a brand-new SUV or boob job after they've been informed about Peak Oil? If the media was to publicly announce the truth about Peak Oil, people would cease consuming, investment in the stock market would evaporate, the economy would plunge, chaos would ensue, and the whole deck of cards would come crashing down before our leaders and corporate elite have a chance to secure their own well-being.

2.   The average American is not emotionally prepared to deal with Peak Oil. Peak Oil is a literal death sentence to much of our population as well as a figurative death sentence to the energy-intensive American way of life. Nobody likes to deal with their own mortality. When faced with such news, most people choose to "kill the messenger." Those of you who have attempted to tell your friends and family the truth about the impending crisis know exactly what I'm talking about.

The harsh truth is we don't want the truth. You know the cliché. I'm an attorney by trade, so I can't help myself. Are you ready? Here goes:

We can't handle the truth!!!

If we could, Fox News wouldn't be the number-one cable news station.

66. How do I know Peak Oil isn't just a myth made up by the oil companies and other corporate interests to drive the price of oil up and keep me enslaved to a scarce resource?

For a few reasons:

1.   If this was all a ploy by the oil companies to drive up the price of oil, you would have been hearing and reading about Peak Oil at least 10 or 15 years ago. By now the mainstream media would be covering this at least as much as they've covered the Kobe Bryant case. It would be the top story on Fox and CNN every

evening and on the front page of *The New York Times* and *Washington Post* every morning.

2. Wall Street and the corporate "powers that be" most definitely do not want you to truly grasp the seriousness of this situation. Upon finding out the truth about the ramifications of Peak Oil, many people pull their money out of the market, trade in their gas guzzler for a bike, attempt to pay off their debts, sell their overpriced home, and minimize their purchase of consumer goods. In other words, they do everything the corporate "powers that be" don't want them to do.

    This is what I find so funny about the various "unlimited oil" theories making the rounds on the Internet lately. If oil was truly an unlimited resource, you can bet your bottom dollar the mainstream media would be shouting it from the rooftops! What better way to keep you consuming and incurring more and more debt than to reassure you there's enough energy to keep the party going forever?

3. If the "greedy oil companies" were as in control of oil prices as many people assume they are, they would have lowered the price of oil from $55/barrel in the weeks leading up to the 2004 election to ensure oil-men Bush and Cheney get back into office. (Note: I'm writing this on October 25, 2004)

67. In light of the energy situation we are facing, why are our leaders spending money and cutting services like there's no tomorrow?

From their perspective, there is no tomorrow. They know that the future will be characterized by conflict, not cooperation. Why bother spending money on higher education when most of today's young people are more likely to be heading to the Middle East than the Ivy League? Why bother spending money on Social Security when the average recipient isn't contributing to the GNP at a time when we need all the money we can get to finance oil wars?

It's not just American leaders who feel this way. The leaders of most of the world are in complete agreement with them. As the president of the World Bank, James Wolfensohn, told an audience in Australia:

> We are spending 20 times the amount on military expenditure than what we are spending on trying to give hope to people. If a Martian came to Earth and read the UN's millennium development goals, and

then looked at what we're doing, she'd think we were mad. We are spending a trillion dollars a year on defense. We've got $350 billion being spent in agricultural tariffs, but we're spending maybe $50 billion on development.[163]

As the data cited by Wolfensohn indicates, the leaders of the world know that the first half of the 21st century is going to be characterized by global resource wars. They are simply taking the appropriate preparations, and making the appropriate budget cuts.

68. Why are we going off to the Moon and then to Mars at a time when we should be dealing with these oil shortages?

The US is going to the Moon and to Mars is for four reasons:

A. Because Halliburton wants to go:

How surprised are you to find out Dick Cheney's old company, Halliburton, has been attempting to position itself to profit from the Mars mission? According to Halliburton scientist Steve Streich:

> Drilling technology for Mars research will be useful for the oil and gas industries. The oil industry is in need of a revolutionary drilling technique that allows quicker and more economical access to oil reserves. A Mars mission presents an unprecedented opportunity to develop that drilling technique and improve our abilities to support oil and gas demands on Earth.[164]

It's possible that Halliburton's excursion into space will lead to the development of some new form of oil-drilling technology. However, doesn't it seem a bit odd to develop drilling technology for a low-gravity environment such as Mars if your hope is to develop drilling technology for a high-gravity environment such as Earth? Wouldn't it be a lot cheaper to develop the technology in Texas or Oklahoma instead of Mars?

Given Halliburton's history of cost overruns on billion-dollar contracts in Iraq, one can only imagine what shenanigans will take place once they are receiving ungodly sums of money to operate top-secret technology that nobody but Halliburton knows how to operate, in an astonishingly harsh environment which is located roughly 300 million miles away from the nearest government auditor.

As far as government contracts go, that's about as good as it gets.

B. To develop and deploy space-based weapons ("*Star Wars Missile Defense*"):

The domination of space has been a goal of the US military for many years. We are returning to space to achieve this goal. As General Joseph Ashy, former commander-in-chief of the US space command, said in 1996:

> It's politically sensitive, but it's going to happen. Some people don't want to hear this, and it sure isn't in vogue, but absolutely we're going to fight in space. We're going to fight from space and we're going to fight into space. That's why the US has developed programs in directed energy and hit-to-kill mechanisms. We will engage terrestrial targets someday — ships, airplanes, and land targets — from space.[165]

In January 2004, Tom Feeney (R-FL) was equally adamant that the US participate in space based warfare:

> Somebody is going to dominate space. When they do, just like when the British dominated the naval part of our globe, established their empire, just like the United States has dominated the air superiority, ultimately, whoever is able to dominate space will be able to control the destiny of the entire Earth.[166]

A month later, the Pentagon released the "US Air Force Transformation Flight Plan." According to a March 15, 2004 *San Francisco Chronicle* article by Theresa Hitchens, the plan explains that the following weapons systems will be developed over the next few years: "an air-launched missile designed to knock satellites out of low orbit, space-based lasers for attacking both missiles and satellites, and 'hypervelocity rod bundles' (nicknamed "Rods from God") which will burst from space and slam into deeply buried bunkers."[167]

The deployment of these weapons will allow the US to better prosecute the war on terror which, as Part VI explains, is really a war for oil. The US's desire to deploy space-based weapons became frantic after its petro-rival, China, sent a man into space.

C. To distract and placate the average American with promises of a new "super-fuel" from space:

As Julie Wakefield explained in a June 2000 article for *Space.com*:

> Researchers and space enthusiasts see helium 3 as the perfect fuel source: extremely potent, nonpolluting, with virtually no radioactive by-product. Proponents claim it's the fuel of the 21st century. The

trouble is, hardly any of it is found on Earth. But there is plenty of it on the moon.[168]

Helium-3 sounds great until you find out that in order to use it we need to:

1. Build a nuclear fusion reactor. We've been trying to do this for 40 years yet we are still 40 years away from it.

2. Strip mine large surfaces of the Moon.[169]

3. Develop massive amounts of new technology. According to a November 15, 2003 article in *The Asia Times* entitled, "An Energy Source that's Out of this World," utilization of Helium-3 would require, "superconducting magnets, plasma control and diagnostics, robotically controlled mining equipment, life-support facilities, rocket-launch vehicles, telecommunications, power electronics, etc."[170]

In other words, Helium-3 is another red herring designed to distract and placate you, just like the "Hydrogen Economy!" The average American, however, is so easily hypnotized by the possibility of techno-messiahs, that when they hear a glowing press release about something as exotic as Helium-3, their critical faculties completely shut down.

Likewise, venture capitalists, who often lack in brain cells what they possess in dollar bills, typically begin drooling when they hear promises that the market for Helium-3 is as large as the market for oil and gas. Blinded by lust, they start tossing dollar bills at the gyrating naked emperor who promises them, "we can invent our way out of this and I will see to it we make the appropriate investments with the right companies."

Folks, Helium-3 not a viable, scalable replacement for fossil fuels for many of the same reasons Hydrogen is not a viable, scalable replacement for fossil fuels. We currently get none of our energy from Helium-3. We have none of the technology necessary to harness it. We have no infrastructure adapted to run on it. We don't even have a miniscule supply of the stuff yet.

The corporate elite are playing with our emotions when it comes to pie-in-the-sky plans like space travel and space-derived fuel sources. They know full well that the prospect of space travel has deep emotional meaning to many of us. They are more than willing to use this fact to dupe us into accepting the latest techno-messiah as our financial and social savior.

I hate to be the bearer of bad news, but you must understand something: we no longer have any business going to space. We originally went to the

Moon in 1969. At that time per-capita energy production was still growing at an astonishing yearly rate. That was then. Peak Oil is now. We have to adjust our goals accordingly.

I know this sounds terribly pessimistic, especially to those of you who (like me) are inspired by the "final frontier," but consider the following analogy: imagine you inherit a huge sum of money shortly after having children. Naturally, you hope to eventually use that money to send your young children to study abroad at the finest universities or to participate in some similarly high-minded and inspiring endeavor. Maybe you even look at college catalogues and dream about the day junior will be heading off to Oxford or Yale or where have you.

Over the next few years, however, you proceed to squander the money on gas-guzzling vehicles, overpriced homes, and useless crap like 50-inch plasma televisions. Consequently, the inheritance is largely depleted by the time your children are in high school.

Unfortunately, you can no longer afford to send your children to study abroad at super-expensive universities! Yale is simply no longer a possibility. You squandered the inheritance. As result your kid is staying put at the local junior college, not jetting off to Paris. It's time for you to be an adult about the situation, take stock of whatever money you have left, and direct it to your family's most pressing needs.

As a society, we have wasted much of our energy (fossil fuel) inheritance on cars, consumer goods, and other useless crap. Now that our inheritance has been largely depleted, it's time for us to grow up and face reality. We need to direct what little inheritance we have left to our most pressing needs, like food and water production, not the fulfillment of our childhood Star Trek fantasies.

Of equal importance, we cannot willingly allow our leaders to play on these fantasies in order to waste the precious little time, money and energy we have left.

D.   To send more US jobs offshore and look for WMD.

Just joking about the jobs thing.

However, as far as WMD: let me remind you that Mars is the planet of war! As George W. Bush has explained time and time again, he has taken the war on terror to Baghdad and Kabul so it doesn't have to be fought in Boston and Kansas. Doesn't it only make sense to next take the war on terror to Mars and the Moon so it doesn't have to be fought in Mississippi and Missouri?

69. What about Bush's plan to give amnesty to the illegal immigrants from Mexico? Does that have anything to do with Peak Oil?

Mexico is the third-leading oil supplier to the US, exporting about 2 million barrels of oil per day to the US. According to Dick Cheney's National Energy Report released in May 2001, "Mexico is a leading and reliable source of imported oil. It has a large reserve base, approximately 25 percent larger than our own proven reserves."[171]

On May 8, 2003, the US Congressional Committee on International Relations voted to tie reform of US immigration laws with a requirement that Mexico open up its state oil company, Petroleos Mexicanos, to US corporate investors.[172] In other words, the US told Mexico, "Give us your oil and we will give you favorable immigration laws."

The plan was hugely unpopular with both Democrats and Republicans. Even fanatical, right-wing talk show hosts were astonished when they heard about Bush's amnesty plan. Bush had a really good reason to go against his "base." We need that oil more than Bush needs the right-wing fanatic vote. That should tell you how badly we need that oil.

70. Does Peak Oil have anything to do with the war on drugs?

Yes.

There is a famous saying which goes, "All 'wars' are about GOD: gold, oil and drugs." The war in Afghanistan is instructive:

In July 2000, Taliban supreme leader Mullah Mohammad Oman imposed a ban on opium production. This act eliminated 70 percent of the world's opium production.[173]

This created a huge problem for the US economy, because it is supported in large part by money from the drug trade. Like any good businessman, a drug lord knows the best place to invest his money is the US stock market. The narcotics racketeers invest as much, if not more, money in the market, towards political campaigns, and even towards benevolent charitable causes as legitimate business interests do.

Estimates are that 2/3 of the profits from the global narcotics trade stays in US banks.[174] While the **Department of Justice** estimates that $100 billion in drug funds are laundered in the US each year, other research places the figure between $250-$500 billion per year.[175]

On average, a bank will loan out $6 for every $1 that it holds in deposits. Likewise, every dollar that is invested in a publicly traded company results in about $6 in circulated wealth due to what is known as the "pop factor."[176]

The ramifications of this are a bit staggering: $500 billion laundered in drug money results in 3 trillion dollars in cash transactions resulting from the drug trade, or about 1/3 of the US GDP in 2002.

Let me restate that so as to be perfectly clear: depending on which figures you believe, 6-33 percent of every dollar you've earned or spent has been the result of the narcotics trade. In order for the US economy to function, we have to keep our children hooked on billions of dollars of illegal narcotics each year.

With that much money at stake, there is little wonder the US was so motivated to invade Afghanistan. Following the invasion, Afghan farmers began replanting opium fields at a furious pace. These fields were usually located in areas controlled by the Northern Alliance, which receives money and arms from the US. By 2002, opium production had returned to pre-2000 levels.[177] By 2003, the production of poppy (used to make heroin) had risen to a level 36 times higher than in the last year of rule by the Taliban.[178] As a result, the world drug market boomed. The US stock market responded accordingly.

Have you ever wondered how the economy could possibly have a "jobless recovery?" It doesn't make common sense. How can the market be doing so well the past couple of years when people have less money to spend on the goods and services sold by the corporations traded on the market?

Here's how: you don't need to sell many goods and services when the world narcotics market is booming and the drug cartels are throwing money at you like it's going out of style.

If this is hard to get your head around, think about it this way: the local business that is actually a front for drug money doesn't need to sell very much of whatever it purports to sell or hire very many employees, does it? Of course not. It just needs to sell enough goods and services and employ enough of a skeleton work force to hide the fact it's laundering drug money.

That's pretty much how the whole US economy works.

The overlap the war on terror and the war on drugs goes much further than just Afghanistan. Many of the South American cocoa fields targeted for "eradication" by the US, for instance, just so happen to sit atop oil fields earmarked for large oil companies.[179]

Many of you may find the degree to which the drug trade controls our economy a bit shocking. You may be tempted to call me a "conspiracy theorist" for pointing out a very uncomfortable truth about our lives. If doing so makes you feel better, have at it, but keep in mind the CIA has admitted the drug cartels are running the country. In 1995 William Colby (the former director of the CIA) stated, "The drug cartels have stretched their tentacles much deeper into our lives than most people believe. It's possible they are calling the shots at all levels of government."[180]

On a semi-related note, when I first read Colby's statement, I couldn't help but to wonder, "Who then was calling the shots on 9/11?"

71. I'm a Baby Boomer. What can I expect in the years to come?

You can expect the evaporation of any entitlement programs such as Social Security and Medicare, in addition to being despised by your grandchildren's generation.

A. *Will Social Security and Medicare still exist in a few years?*

No.

If Alan Greenspan's announcement in March 2004 did not make it perfectly clear, let me do so: **you can kiss Social Security and Medicare goodbye.** The numbers just don't add up. As economist Dr. John Attarian explained in a summer 2002 article entitled, "The Coming End of Cheap Oil:"

> The Congressional Budget Office projects that spending for Social Security, Medicare, and Medicaid will rise from 7.8 percent of GDP in 2001 to 14.7 percent by 2030. Assuming federal revenues remain roughly 20 percent of GDP, and entitlement programs are unchanged, the General Accounting Office forecasts that by 2030 federal outlays will be roughly 28 percent of GDP and Social Security, Medicare, Medicaid and interest on the debt will take 75 percent of revenues. By 2050, outlays will be almost 40 percent of GDP, and revenues will cover only half of them, making the budget deficit some 20 percent of GDP.[181]

As Dr. Attarian goes onto explain, these terrifying calculations do not even take into account the devastating effects of the coming oil shocks. Nor do they account for the effects of global climate change, large-scale international warfare, terrorism and the crushing debt load carried by both the US government and most of its citizens to the list as well.

Sorry Boomers, but our leaders are looting your entitlement programs to finance tax cuts and oil wars.

B. *What will our grandchildren think about us?*

If you're a Baby Boomer, your parents' generation is often referred to as "The Greatest Generation." Unfortunately, your generation is likely to be referred to as the "The Greatest Wasting Generation."

To understand why, imagine you are a member of your grandchildren's generation, born in the year 2000. You will turn 16 in the year 2016, just as society is collapsing. You get your driver's license, but due to the worsening state of the economy and the incredibly high cost of gas, you are unable to get a car. Your prospects for college are virtually nonexistent, not because of any academic shortcomings, but because there is virtually no financial aid available and many colleges have closed. You have seen many of your friends drafted into the latest oil war, and are anxiously trying to figure out a way to avoid their fate. As you contemplate your future, or lack thereof, you look around and see evidence of decades of wasteful consumption. Justifiably angry, you look for a scapegoat. Who better to scapegoat than the generation that consumed the most, conserved the least, paid little attention to the true actions of their government, and refused to address the ramifications of their behavior until it was too late?

As author Daniel Quinn has stated:

> If we consume the world until there's no more to consume, then there's going to come a day, sure as hell, when our children or their children are going to look back on us — on you and me — and say to themselves. "My God, what kind of monsters were these people?"[182]

72. I'm a member of Generation X or Generation Y. How will Peak Oil affect me in particular?

If you haven't already figured it out, you can forget about the "house with a picket fence and a Lexus in the garage" dream. Three areas that are likely to be of interest to you are: your debts, your education, your investments, and your chance of being drafted.

A. *What's going to happen to people in debt?*

Big business is already lobbying the government to slowly reinstitute a feudal style system of slavery/indentured servitude here on American soil.

The future will not be a good time to be in debt. Our prison population has grown by approximately 500 percent since 1980. We now incarcerate 1 out of every 75 men in the US at any given time. About 1 out of 20 men can expect to be incarcerated at some point during their life. The debt loads carried by many Americans have skyrocketed during that same time. As the economy begins to dissolve, and unemployment becomes endemic, many people will not be able to pay their debts. Although the practice of debtors' prison was abolished centuries ago, it could again become viable for the following reasons:

1. People unable to pay their debts will receive the opportunity to do so.

2. Industry will benefit as it will be able to outsource labor to domestic work camps.

3. The construction and maintenance of work camps will provide additional jobs.

If the prospect of debtors' prison seems outrageous to you, consider the fact that many states punish low-level drug transactions with 10-, 15-, and even 25-year-to-life prison sentences. Simple drug possession is now the number-one crime for which people are incarcerated.

If the government is willing to incarcerate an individual for 25 years as punishment for a $20 drug transaction, what makes you think they will hesitate to incarcerate an individual for a $20,000 bankruptcy, especially when that person will provide a source of cheap labor (energy) to a crumbling economy?

If you find this hard to believe, I ask you: where did big business get labor/energy prior to the advent of fossil fuel-powered machinery?

Now that the supply of cheap fossil fuel energy is diminishing, big business will have no choice but to return to a **system of slavery** to maintain its profits.

Legally, big business is practically required to take this course of action. Remember, the corporate officers of any publicly traded corporation have a fiduciary duty to do what is best for the company, so long as their actions are not illegal.

There is nothing illegal about lobbying the government to slowly bring back a system of labor resembling slavery and then making use of this system once it's in place.

It's relatively easy for these corporations to effectively lobby government officials as these government officials, more often than not, used to work for the same corporations now lobbying them.

This will work out great for the big business interests that both George W. Bush and John F. Kerry have courted throughout their presidential campaigns. Regardless of who is installed in the White House come January 2005, these contractors will have access to an ever-increasing pool of cheap, domestic labor.

B. What's going to happen to my investments?

Assuming the stock market even still exists in the year 2015, your investments will evaporate as the Baby Boomers attempt (in vain) to pull their money out for retirement and the entire modern system of economics collapses due to massive shortages of energy. Even Warren Buffet has pulled out of the market and has most of his money in various foreign currencies. That should tell you something.

C. Will I be drafted to fight for oil?

If you were born after 1980, you may be drafted to fight for oil. See Part VII for more information.

73. I own a Hummer. What can I expect in the years to come?

You're going to have a very difficult time obtaining gas, and not because of the cost. When people are sitting in three-hour-long gas lines, they're going to look for a scapegoat. As the owner of the most infamous gas-guzzling SUV ever built, you're a natural target. You may want to get that thing bulletproofed before gas rationing starts.

74. I'm having trouble believing that a country as powerful as the United States is on the verge of collapse.

Let's look at what has happened to the US in just the past four years: World Trade Center destroyed, budget surplus vanished, affordable health care gone, honest elections gone, three million jobs gone, hundreds of publicly traded companies gone bankrupt, Social Security close to gone, government oversight of big business gone, weakened infrastructure, shrinking middle class, botched invasions based on lies have turned into tragic quagmires, undermined civil liberties, and despite a national security apparatus equipped with the best technology money can buy in addition to technology money can't buy, we can't find the one Arab guy in the whole world who is both 6'6" and hooked up to a dialysis machine 24/7.

This is what happens when any civilization overshoots its resource base. It degenerates into chaos, barbarism, and corruption. It isn't a new thing. We won't be the first superpower to collapse. Over the course of history, the collapse of civilizations has been as inevitable as death and taxes. Any good book on the fall of the Roman Empire will give you a case of deja- vu next time you watch the evening news.

Those of us lucky enough to live in the US are like the cool kids who got invited to the big party. Unfortunately, as Richard Heinberg has said, "the party's over."

75. Is it possible that the government is actually trying to speed up the collapse?

Yes.

From the government's perspective, a fast collapse may be better than a slow one. A slow crash may simply exacerbate the problems, because the population at the turning point of oil production will be even larger than it would be at an earlier date. The higher the population is, the higher the number of deaths that will result when the cheap oil runs out. In the eyes of our government, a fast crash may be the "kindler, gentler" alternative. It also gives the American public less time to wake up as to what is really going on.

This would certainly explain why the government gives tax breaks to SUV owners at a time when it should be encouraging conservation. It would also explain why the deficit is being run up to a level that virtually assures the government will be bankrupt by the year 2011, which just happens to be the same year many predict world oil production will peak. The chance of this being a coincidence is slim. It appears to be a deliberate manipulation to squeeze out every dollar from those not in the know before it's too late.

Part VI. Peak Oil and America's March Towards a Fascist-Feudal State

"Beware of the military-industrial complex."
-Dwight D. Eisenhower

"I never would have agreed to the formulation of the Central Intelligence Agency back in '47, if I had known it would become the American Gestapo."
-Harry S. Truman

"Fascism should more properly be called corporatism because it is the merger of state and corporate power."
-Benito Mussolini

"The individual is handicapped by coming face-to-face with a conspiracy so monstrous he cannot believe it exists."
-J.Edgar Hoover

"Let us never tolerate outrageous conspiracy theories concerning the attacks of September the 11th."
-George W. Bush

"God told me to strike at Al Qaeda and I struck them. And then he instructed me to strike at Saddam, which I did. With the might of God on our side we will triumph"
-George W. Bush

"I believe that I am acting in accordance with the will of the Almighty Creator."
-Adolf Hitler

"Terrorism is the best political weapon, for nothing drives people harder than a fear of sudden death."
-Adolf Hitler

"Never forget that everything Hitler did in Germany was legal."
-Dr. Martin Luther King

76. Does Peak Oil have anything to do with September 11th?

Absolutely.

The standard story regarding 9/11 is that Osama Bin Laden and his followers were angry at the US because we have military bases located near Muslim holy sites in Saudi Arabia. Motivated by this anger, Bin Laden and his followers callously attacked the US.

Variations of this story are as numerous as the stars. Many people believe the various national security agencies simply failed to properly coordinate their anti-terrorism efforts. Others think elements of the government knew of the attacks beforehand and allowed them to happen.

I want you to consider an explanation far more heinous than any you may have come across yet: that elements of the US government actually orchestrated the attacks so as to gain domestic support for two wars: a worldwide war for oil, better known as the war on terror, and a domestic war on dissent.

An in-depth analysis of our government's role in the attacks is really beyond the scope of this book. Ultimately, it's not necessary for me to convince you of US government complicity in the attacks to convince you we have a full-blown meltdown of petrochemical civilization on our hands. However, since 9/11 and Peak Oil are so intertwined, I cannot simply skip the issue altogether.

What I hope to accomplish in the next few pages is to give you a brief overview of what I feel are a few of the most compelling pieces of evidence for US government complicity in the attacks. I'm not trying to prove it beyond a reasonable doubt as doing so would require all of the following:

1. Tremendous investigative abilities;

2. A set of brass balls;

3. A complete absence of a fear of bullets heading in one's direction.

I posses none of these things. Luckily there are authors such as Michael Ruppert and David Ray Griffin who do. If the following pieces of evidence pique your interest, I recommend you check out Ruppert's forthcoming book, *Crossing the Rubicon: The Decline of the American Empire at the End of the Age of Oil* or Griffin's book, *The New Pearl Harbor: Disturbing Questions About the Bush Administration and 9/11.*

A. **Do you have documented, irrefutable proof the US government is capable of conspiring to plot a 9/11-style attack against its own citizens?**

Yes.

On March 13, 1962, a document known as Operation Northwoods was presented to President John F. Kennedy. The document was declassified in 1997. By 2001, both the *Baltimore Sun* and *ABC News* had run stories on the document.[183] The contents of the document are quite sobering, and I encourage you to read it in its entirety, for yourself. The document is available from dozens of reputable online sources such as the National Security Archive located at George Washington University.[184] It is thoroughly analyzed by author James Bamford in his 2001 book *Body of Secrets.*

The goal of Operation Northwoods was to get public support for an invasion of Cuba. The Joint Chiefs of Staff believed (rightfully so) that Americans would only support a war against Cuba following a series of surprise attacks by Cuban "terrorists" against the US.

The Northwoods document frequently refers to staging **fake terror attacks against American citizens.**

Many people believe such a "conspiracy" could never be pulled off by the US government because it would be impossible to keep it a secret. This belief ignores two facts:

1. It makes little sense to believe a group of Arab dissidents are capable of keeping such an operation a secret while simultaneously believing the government agencies that trained them aren't.

2. As the Northwoods document clearly demonstrates, the Joint Chiefs of Staff don't share this belief. There is likely a good reason they don't: they've been around the block a few more times than the average American who believes, "the government couldn't keep such a plan a secret." The document explains that:

> Such a plan would enable a logical build-up of incidents to be combined with other seemingly unrelated events to camouflage the ultimate objective and create the necessary impression of Cuban rashness and irresponsibility on a large scale.

> The plan would also properly integrate and time phase the courses of action to be pursued. The desired resultant from the execution of this plan would be to place the United

States in the apparent position of suffering defensible grievances from a rash and irresponsible government of Cuba and to develop an international image of a Cuban threat to peace in the Western Hemisphere.

If I didn't know better, I'd say that paragraph plainly indicates the Joint Chiefs were plotting a "conspiracy."

The specific recommendations made in the Northwoods document are as eye-opening as they are sickening:[185]

1. Stage mock attacks, sabotages and riots and blame it on Cuban forces.

2. Sink an American ship at the Guantanamo Bay American military base or destroy American aircraft and blame it on Cuban forces.

3. Harassment of civil air, attacks on surface shipping, and destruction of US military drone aircraft by MIG-type planes would be useful as complementary actions.

4. Destroy a fake commercial aircraft supposedly full of "college students off on a holiday."

5. Stage a "terror campaign," including the "real or simulated" sinking of Cuban refugees. The document states:

> We could develop a Communist Cuban terror campaign in the Miami area, in other Florida cities and even in Washington. The terror campaign could be pointed at Cuban refugees seeking haven in the United States. We could sink a boatload of Cubans en route to Florida. We could foster attempts on lives of Cuban refugees in the United States even to the extent of wounding in instances to be widely publicize.

Fortunately, Kennedy rejected the plan as outrageous. Between his rejection of Northwoods and his hope to reform the monetary system, there is little wonder he was assassinated

To those of you who claim the idea of government complicity in the attacks to be "paranoid conspiracy," I insist you get yourself a copy of the Northwoods Document for yourself. Its existence alone moves the possibility of the US government staging fake terror attacks against its own citizens out of the realm of "paranoid conspiracy that can be quickly dismissed," and into the realm of "conceivable possibility that must be discussed."

Read through the Northwoods document and simply replace "Cuba" with "Iraq," "Afghanistan," or "Al-Qaeda" and you've got a blueprint for 9/11.

They planned to stage fake terror attacks against us in 1962. What's to keep them from staging fake attacks nearly 40 years later when they have so much more technology with which to orchestrate the attacks and are so much more motivated to get our support for the oil-wars America will need to fight to maintain access to the hydrocarbon energy it requires?

B. Prior to 9/11, did any of our leaders indicate that a "new Pearl Harbor" might accelerate the accomplishment of their goals?

Yes.

The Project for the New American Century, or PNAC, is a neoconservative Washington-based think tank created in 1997. Dick Cheney, Donald Rumsfeld, Paul Wolfowitz and Defense Policy Board chairman Richard Perle are founding members of PNAC.

In September 2000, the PNAC released a report entitled: "Rebuilding America's Defenses: Strategies, Forces, and Resources for a New Century," which called for or predicted the following:

1. Massive hikes in military spending;

2. The establishment of American military bases in Iraq and Saudi Arabia;

3. US willingness to violate international treaties;

4. US control of the world's energy sources;

5. US development of space-based weapons;

6. "Advanced forms of biological warfare that can **target specific genotypes** may transform biological warfare from the realm of terror to a politically useful tool;" (Emphasis added)

7. US willingness to use nuclear weapons to accomplish its goals;

8. Possible conflict with China.

The proposal is available, in its entirety, on the PNAC Website,

107

www.newamericancentury.org. I encourage you to read it for yourself, but not until after you've had a good stiff drink and are sitting down.

The folks at PNAC may be full-blown nutcases, but they aren't complete idiots. They realized the American people would never support their outrageous goals unless we were thoroughly traumatized first. Thus, on page 52 of the proposal, PNAC proposes, "Further, the process of transformation, even if it brings revolutionary change, is likely to be a long one, absent some catastrophic and catalyzing event – like a new Pearl Harbor."

On the morning of 9/11, a "new Pearl Harbor" occurred. The event was certainly catastrophic and it certainly catalyzed the American people into supporting the absolutely outrageous level of defense spending and hyper-militarized foreign policy advocated by PNAC.

When investigating a murder, the first question any good investigator asks is "Who would benefit from this?" The answer to that question usually provides you with your first suspect.

C. Did the attacks get the American people to support invasions they likely never would have supported had the attacks not happened?

Yes.

The US government had sought to control the oil located under the Caspian Sea for years prior to 9/11. In order to control that oil, a pipeline needed to be built through Afghanistan.

In 1995, Petroconsultants shocked the oil industry with their report entitled, "World Oil Production, 1950-2050." The report predicted global oil production would peak around the year 2000 and decline by as much as 50 percent by 2025. (See also question number one, "When Will Peak Oil Occur?")

The imminence of the oil peak and the severity of the decline as predicted by Petroconsultants made construction of a pipeline through Afghanistan an urgent priority for both the oil companies and the US government. That very same year, Unocal invited some of the leaders of the Taliban to Houston, where they were entertained like respected dignitaries. Negotiations for the construction of the pipeline were initiated and continued until as late as August 2001.[186]

The US government's relationship with the Taliban was as cuddly as Unocal's. In 1997 a US diplomat told author Ahmed Rashid "the Taliban will probably develop like the Saudis did. There will be Aramco [the former US oil consortium in Saudi Arabia] pipelines, an emir, no parliament and lots of Sharia law. We can live with that."[187]

That Bill Clinton had his roving eye on the oil fields of the Caspian Sea is evidenced by a speech he gave in Azerbaijan in 1997 in which he stated "In a world of growing energy demand our nation cannot afford to rely on any single region for energy supplies. By tapping the Caspian Sea resources, we diversify our energy supply and strengthen our nation's security."[188] Clinton's sentiments were later echoed by other high-level sources, such as the following article from the Foreign Military Studies Office of Fort Leavenworth, which was published three months prior to the 9/11 attacks. The article states:

> The Caspian Sea appears to be sitting on yet another sea — a sea of hydrocarbon. Western oilmen flocking to the area have signed multibillion-dollar deals. US firms are well represented in the negotiations, and where US business goes, US national interests follow. The presence of oil resources and the possibility of their export raise concerns for the US.[189]

US negotiations with the Taliban were later broken off, reportedly because the Taliban wanted too much money. Then 9/11 happened, the US invaded Afghanistan, former Unocal employee Harmid Karzai was installed as the President of Afghanistan, and the pipeline project got back underway.[190]

As far as that pipeline was concerned, 9/11 was darn convenient, wouldn't you say?

The American people never would have supported the invasion of Afghanistan had it not been for 9/11. The US was motivated to kill 3,000 of its own citizens because it needed the American people to support the invasion of Afghanistan so that the Unocal pipeline project could get under way. The pipeline would have given the US access to the oil reserves under the Caspian Sea. If these reserves had turned out to be as abundant as we thought they were, Peak Oil would have been delayed for years, if not decades.

D. Is there any evidence of foreknowledge of the attacks?

Yes.

The two airlines involved in the attack were United Airlines and American Airlines.

Between September 6-7, 2001, 4,744 put options were purchased on United Airlines stock. In contrast, only 396 call options were purchased[191] Put options are a speculation that a stock will fall in value. Call options are a speculation that a stock will rise in value.

Curiously, as Michael Ruppert has pointed out, many of the United put options were purchased through the firm Deutschebank/A.B. Brown, which was previously managed by the former executive director of the CIA, A.B. "Buzzy" Krongard.[192]

On September 10, 2001, American Airlines got the same treatment: 4,516 put options were purchased in contrast to only 748 call options.[193]

United and American were the only two airlines whose stock experienced such highly abnormal trading in the days leading up to the attacks. An analysis of the trades leads to only one conclusion: somebody knew that something extremely bad was about to happen to those two airlines and decided they were going to make a little profit from it.

As Tom Flocco explained in a December 6, 2001 FTW article entitled "Profits of Death: Insider Trading and 9/11," the CIA no doubt noticed this bizarre trading since they monitor stock trading in real time using highly sophisticated computer programs.[194]

The identities of the investors who made the trades is still unknown to the general public as the $2.5 million in profits produced by the trades went uncollected.[195] Given the vast investigative resources of the CIA/FBI/SEC, in addition to the technology the agencies employ to monitor stock trading, it is hard to imagine they don't know who the 9/11 profiteers are.

These bizarre trading patterns were reported by several mainstream media sources (such as the *San Francisco Chronicle)* in the weeks following the attacks, but have been completely ignored ever since. At the very least, you should be suspicious as to why the 9/11 insider trading issue hasn't received consistent coverage on the nightly news while the Martha Stewart, Scott Peterson, Kobe Bryant, and Michael Jackson cases have been covered virtually 24/7. Clearly somebody had foreknowledge of the attacks or they would not have been conducting such abnormal trades. The fact that the mainstream media has ignored such a bombshell story (which would generate tons of controversy, which would generate ratings) is extremely suspicious in and of itself.

110

E. Did anything happen on the morning of 9/11 that tends to show our government was negligent or complicit in the attacks?

Yes.

Issues revolving around the scrambling of fighter jets the morning of 9/11 tend to show elements of the US government were at best negligent in responding to the attacks, and at worst complicit in the attacks.

The official policy of NORAD and the FAA is to scramble fighter jets the moment any airplane veers off its flight path by even a minor degree. The scrambling of the jets requires absolutely no input from the President or anybody else. It is an automatic, routine, and well-practiced exercise that was carried out on 67 occasions between September 2000 and June 2001.[196] One of these occasions was even covered in *Sports Illustrated* because it involved the private jet of pro-golfer Payne Stewart.[197]

The contrast between the events surrounding the Stewart incident and the events of 9/11 are illuminating. When Payne's small, private jet deviated only slightly from its flight plan, the FAA/NORAD acted as follows:

1. Scrambled fighters immediately;

2. Once scrambled, the fighters proceeded to Payne's jet at full speed.

In contrast, when four large passenger jets were simultaneously hijacked and taken radically off their flight plans on 9/11, the FAA/NORAD acted as follows:

1. Waited 75 minutes to scramble aircraft;

2. Once scrambled, the fighters proceeded at one quarter of their top speed.[198]

Allow me to analogize for a moment: let's say the police in your city got word that the most heinous crime in American history was occurring on their watch. Would you be a bit suspicious if you later find out that:

1. They didn't dispatch the SWAT team immediately as their regulations demand that they do? Instead, they waited almost an hour and half before doing so;

2. Once the SWAT team was dispatched, they drove to the crime scene at a leisurely 25 mph?

111

Not only would you be suspicious, you would be outraged! Why then aren't you similarly suspicious and outraged at NORAD and the FAA's failure to scramble those fighter jets?

Given the fact it was well known that four airliners had been simultaneously hijacked the morning of 9/11, the failure to scramble fighter jets tends to show gross criminal negligence, if not out-and-out complicity, on the part of our government in regards to the attacks.

F. Why would they do this to us?

By the late 1990s, the financial elite concluded that if they didn't gain access to more of the world's oil supply, the US economy would dissolve. Motivated in equal parts by survival and greed, they sponsored a coup, also known as the 2000 election.

The elite then went about installing its preferred administration (in this case, the more willingly aggressive one) into power. Once in power, this administration began implementing the fascist agenda of its corporatist benefactors. "What would have to happen to get the American people to accept a radically militaristic agenda abroad and a police state at home?" they asked themselves. They dusted off Operation Northwoods, updated it for the 21st century, and proceeded to orchestrate the 9/11 attacks.

With the American people in shock at the horror of the attacks, the administration declared two wars: a foreign war for oil, a.k.a. the "war on terror," and a domestic war on dissent, a.k.a. "homeland security." The first objective in the war for oil was to secure access to the oil and gas located in the Caspian Sea. This required an invasion of Afghanistan, which was promptly conducted. When the oil find in the Caspian Sea turned out to be an oil bust, the administration quickly turned its attention towards Iraq, which found itself a target in both the oil war and a currency war, which is explained further in Part VII.

G. Again, who would benefit from the attacks?

Even if you are unwilling to believe elements of the US government orchestrated the attacks, there is no doubt about who benefited from their aftermath. As a result of 9/11, Bush's popularity skyrocketed. This allowed him to force the Patriot Act through Congress, pass staggeringly unfair tax cuts, run the deficit up to mind-boggling levels, invade one country thought to sit near significant amounts of oil, invade another country known to sit atop significant amounts of oil, threaten to invade any other country that sits atop or near significant amounts of oil, all while evading scrutiny from the media because few reporters or news agencies have the intestinal fortitude to stand up to the government and/or risk losing ad revenue from the petrochemical industries profiting from Bush's post-9/11 policies.

Similarly, the subsequently declared "war on terror" has been a series of lottery tickets for big business interests such as the defense and pharmaceutical industries. Since 9/11, defense companies have found themselves awash in cash by providing arms and services to the US military and domestic security agencies. Meanwhile, the pharmaceutical industry stands to profit handsomely from proposed government programs that would force citizens to accept vaccinations while insulating the vaccine makers from possible charges of negligence.[199]

If you were investigating an arson case in which the victim:

1. Collected a huge insurance buyout as a result of the fire;

2. Had previously conspired to burn down his house to collect on the insurance;

3. Had stated in writing that it would take a long time to accomplish his incredibly cutthroat business goals in the absence of an event like, "a giant house fire;"

4. Claimed that a bunch of gang members burned down the house because they "hate what he stands for,"

Would you not see through the bullshit?

77. Does Peak Oil have anything to do with legislation such as the Patriot Act?

When the cost of food soars, the military draft is reinstituted, Social Security officially dissolves, gas hits $7.00 a gallon, the stock market crashes, and returning veterans are denied the health care that was promised to them, large-scale rioting will erupt. The only way to control the population will be through the institution of a fascist-style police state. The Patriot Act and related legislation are the foundation for that state.

If you haven't read up on the Patriot Act, you should. Some of its provisions are truly frightening. According to the Act, the government may:

1. Search and seize Americans' papers and effects without probable cause to assist in terror investigations.

2. Imprison Americans indefinitely without a trial.

113

3. Monitor religious and political institutions without suspecting criminal activity.

4. Conduct closed once-public immigration hearings, secretly detain hundreds of people without charges, and encourage bureaucrats to resist public records requests.

Section 802(a)(5) of the Patriot Act defines "Domestic Terrorism" as "activities that... involve acts that are a violation of the criminal laws of the United States or of any state and appear to be intended to influence the policy of a government by intimidation or coercion." According to this definition, Rosa Parks would have been considered an "enemy combatant" for not giving up her seat on the bus. Once designated as such, the government could have legally incarcerated her in Guantanamo Bay.

It's not just the Patriot Act you need to read up on. Few people realize the draconian measures the President can institute via executive orders. A sampling:

- 10995: Right to seize all communications media in the United States.

- 10997: Right to seize all electric power, fuels and minerals, both public and private.

- 10999: Right to seize all means of transportation, including personal vehicles of any kind and total control of highways, seaports and waterways.

- 11000: Right to seize any and all American people and divide up families in order to create work forces to be transferred to any place the government sees fit.

- 11001: Right to seize all health, education and welfare facilities, both public and private.

- 11002: Right to force registration of all men, women and children in the United States.

- 11003: Right to seize all air space, airports and aircraft.

- 11004: Right to seize all housing and finance authorities in order to establish "Relocation Designated Areas" and to force abandonment of areas classified as "unsafe."

- 11005: Right to seize all railroads, inland waterways, and storage facilities, both public and private.

- 11921: Right to establish government control of wages and salaries, credit and the flow of money in US financial institutions.

The president can invoke these executive orders in a number of emergencies, including an **economic crisis.**

If terrorists attack the oil infrastructure in Saudi Arabia, the price of oil could hit $100 practically overnight. This would pull the rug right out from under our highly leveraged and indebted economy. An economic crisis would erupt, giving the President the right to make use of these executive orders.

78. Gosh, don't you think you're making a big deal out of nothing? This is just unnecessary alarmism.

Not according to the following individuals, all of whom either survived fascist police-states or thwarted fascist power-grabs:

A. Professor Milton Mayer

German Professor Milton Mayer lived through the Nazi era. In his book, *They Thought They Were Free: The Germans 1939-1945,* he explains that fascist police states tend to emerge so slowly that the average citizen doesn't realize he's living in a fascist state until its too late:

> You speak privately to your colleagues, but what do they say? They say, "It's not so bad" or "You're seeing things" or "You're an alarmist." And you are an alarmist.

> Each step was so small, so inconsequential, so well explained, or, on occasion, "regretted," that, unless one were detached from the whole process from the beginning, unless one understood what the whole thing was in principle, what all these "little measures" that no "patriotic German" could resent must some day lead to, one no more saw it developing from day to day than a farmer in his field sees the corn growing. One day it is over his head.[200]

Legislation such as the Patriot Act and recent statements made by Attorney General John Ashcroft would likely set off alarm bells from somebody like Mayer. For example, in December 2001, Ashcroft defended the Patriot Act by stating:

115

To those who scare peace-loving people with phantoms of lost liberty, my message is this: your tactics only aid terrorists, for they erode our national unity and diminish our resolve. They give ammunition to America's enemies and pause to America's friends. They encourage people of good will to remain silent in the face of evil."[201]

Ashcroft's testimony before Congress on behalf of the Act was equally scary. While testifying, he stated "...those who oppose us are providing aid and comfort to the enemy."[202]

The one thing you learn in law school is to use precise language. Ashcroft used those specific words for a reason: the crime of treason is defined as "giving aid and comfort to the enemy." Ashcroft would not have used the specific words that define treason to characterize opposition to the government if he wasn't hoping to set a precedent or build a foundation (either for himself or a future Attorney General) for eventually charging those who oppose the government with the crime of treason.

Ashcroft's statements defending the passage of the Patriot Act in 2001 bare an eerie resemblance to the statements made by Adolf Hitler when he defended the creation of the Gestapo in 1934, "An evil exists that threatens every man, woman, and child of this great nation. We must take steps to ensure our domestic security and protect our homeland."

As if Ashcroft's testimony before Congress wasn't disturbing enough, the *Los Angeles Times* reported in 2002 that he had announced his desire for camps for US citizens he deems to be "enemy combatants."[203]

Interestingly enough, Ashcroft's biggest campaign donors to his failed run for Senate were Exxon Mobil and BP Amoco.[204] He understands that the end of cheap oil is the end of the American way of life. He is simply making the appropriate preparations.

B. Vice-President Henry Wallace:

To you "anti-alarmists," please remember: two individuals need not be identical twins to be cousins. Just because a bunch of swastika waving, brown-shirt wearing Nazis aren't marching down the street in your neighborhood doesn't mean the corporatists haven't gained full control of the government. As FDR's second Vice President Henry Wallace warned in the April 9, 1944, edition of *The New York Times*:

> The dangerous American fascist is the man who wants to do in the United States in an American way what Hitler did in Germany in a Prussian way. They claim to be super-patriots, but they would

destroy every liberty guaranteed by the Constitution. They demand free enterprise, but are the spokesmen for monopoly and vested interest. Their final objective toward which all their deceit is directed is to capture political power so that, using the power of the state and the power of the market simultaneously, they may keep the common man in eternal subjection.[205]

Vice President Wallace was trying to warn us not to expect American fascists to behave **exactly** like German fascists. For instance, whereas the Germans allowed themselves to be hypnotized by hyper-patriotic military marching songs, us Americans have allowed ourselves to be hypnotized by hyper-patriotic country music songs.

Those who cry, "Alarmism!" at reading this are likely making the deadly mistake of assuming American fascism will look, feel, and act exactly like German fascism. This assumption is hugely ignorant. Just because you don't look exactly like your cousin doesn't mean the two of aren't closely related.

Regardless of the country it springs up in, fascism is fundamentally about corporations gaining the full reigns of government. Mussolini explained this with his famous quote, "fascism should more properly be called corporatism since it is the merger of state and corporate power." As Henry Wallace pointed out, the goal of American style fascism is the same as the goal of German and Italian style fascism: the "eternal subjection of the common man." The means may not be **exactly** the same, but the ends are.

C. Major General Smedley Butler

FDR's VP had good reason to warn us about American fascism as had it not been for Major General Smedley Butler, a fascist plot against FDR likely would have succeeded.

General Butler is best known for his classic book *War is a Racket* and, in particular, the following quotation (excerpted):

> War is just a racket. A racket is best described, I believe, as something that is not what it seems to the majority of people. Only a small inside group knows what it is about. It is conducted for the benefit of the very few at the expense of the masses.

> The trouble with America is that when the dollar only earns 6 percent over here, then it gets restless and goes overseas to get 100 percent. Then the flag follows the dollar and the soldiers follow the flag.

> I spent thirty-three years and four months in active military service as a member of this country's most agile military force, the Marine

117

Corps. And during that period, I spent most of my time being a high class muscle-man for Big Business, for Wall Street and for the Bankers. In short, I was a racketeer, a gangster for capitalism.

Looking back on it, I felt I might have given Al Capone a few hints. The best he could do was to operate his racket in three city districts. We Marines operated on three continents.

Make no mistake, that's not a quote from some pony-tail sporting pacifist-hippie. General Butler was awarded the Congressional Medal of Honor on two occasions and is widely considered one of the greatest generals in American military history.

Butler's exploits during his 33 years in the Marine Corps are extremely well-known and documented among military historians. What's less well known, or at least less talked about, is that General Butler thwarted a fascist overthrow of the US government a few years after retiring from the Marines. The facts of the attempt have major implications for the events of the past few years.

In 1934 Butler went to Congress with a tale of conspiracy among wealthy American corporatists more terrifying than anything you might hear today on late-night talk radio, watch in a Hollywood movie, or read about on various "fringe" internet forums. Given Butler's background and reputation, Congress could not ignore his claims when he blew the whistle on the corporatist plot. A Congressional investigation was launched, and Butler testified before Congress regarding the conspiracy. According to Butler:

1. In 1933 bond trader Gerald Macquire approached Butler claiming to represent Wall Street broker Grayson Murphy, Singer sewing machine heir Robert Sterling Clark, and other wealthy men with ties to companies such as General Motors and banks such as JP Morgan. The men expressed concern that veterans of World War I might not properly receive the bonuses that had been promised them.[206]

2. Given their immense wealth, Butler found the professed motivations of these men highly suspect. He told them, more or less, to get lost.[207]

3. Later, these men approached Butler again. This time they abandoned any pretense of caring about the veterans' bonuses and explained their real goal was to "protect President Roosevelt from other plotters" by installing a "secretary of general welfare."[208]

118

In order to depose Roosevelt, they needed Butler to lead a march of 500,000 veterans on the White House. The men would be provided with arms and ammunition by the Remington company, which was owned by Irenne Du Pont, one of the plotters.

4. Reportedly Macquire told Butler:

> You know the American people will swallow that. We have got the newspapers. We will start a campaign that the President's health is failing. Everyone can tell that by looking at him, and the dumb American people will fall for it in a second..."[209]

Butler responded to their overtures with the following:

> Yes, and then you will put somebody in there you can run; is that the idea?

> If you get these 500,000 soldiers advocating anything smelling of Fascism, I am going to get 500,000 more and lick the hell out of you, and we will have a real war right at home.[210]

After Butler came forward, Congress formed the McCormick-Dickstein Committee (the precursor to the infamous House on Un-American Activities) to investigate the matter. They listened to Butler's testimony in a secret session that met in New York City on November 20, 1934. On February 15, 1935 the committee released its preliminary findings:

> In the last few weeks of the committee's official life it received evidence showing that certain persons had made an attempt to establish a fascist organization in this country. There is no question that these attempts were discussed, were planned, and might have been placed in execution when and if the financial backers deemed it expedient . . . your committee was able to verify all the pertinent statements made by General Butler, with the exception of the direct statement suggesting the creation of the organization.

> This, however, was corroborated in the correspondence of MacGuire with his principal, Robert Sterling Clark, of New York City, while MacGuire was abroad studying the various forms of veterans' organizations of Fascist character.[211]

The committee later released another report in which it downplayed the plot and deleted much of Butler's testimony. Butler went on national radio to

119

denounce the second report, but the story was largely ignored by the mainstream media except for a few occasions in which Butler was ridiculed.

In 1971, former Speaker of the House John McCormack stated:

> If General Butler had not been the patriot that he was, and if they the plotters had been able to maintain their secrecy, the plot certainly might very well have succeeded, having in mind the conditions existing at the time. . . . If the plotters had gotten rid of Roosevelt, there is no telling what might have taken place.[212]

Not for our generation. We can simply look around to see what would have happened. Had General Butler failed, we would likely have seen the events of the past few years simply transpire two generations earlier. The goal of the fascists who sought to overthrow the government in the 1930s have been completely, albeit gradually, achieved over the past two generations. Much to the dismay of General Butler, the US military is now used as tool to line the pockets of big business interests while the US border is left largely unsecured.

D. Conclusion

As people like Mayer, Wallace, and Butler have tried to warn us, "it can happen here."

By the logic of our corporatist-hijacked government, the institution of a fascist-style police state is entirely reasonable and in fact, now necessary given the economic anarchy likely to accompany declining oil production. The needs of the people must be subordinated to the needs of their corporatist controllers – increased and unfettered access to the world's oil supply.

Finally, let us not forget that a fascist-style police state geared towards maintaining an abundant energy supply and continued economic growth existed for hundreds of years on American soil.

It was called slavery.

79. Would this explain why our military and police forces have been investing so much in "crowd control" and "non-lethal" technology?

Yes.

You didn't see the Marines flushing Uday and Qusay out of that house with anything non-lethal, did you? You think the Marines are going to bust into Osama's cave with non-lethal weapons? No way; they're saving that stuff for the folks who will be rioting when gas hits $7.00 per gallon and unemployment hits fifty percent.

80. If he's elected president, will John Kerry's foreign and domestic policies be much different than George W. Bush's?

Probably not.

Like Bush, Kerry is extremely wealthy, a Yale graduate, a member of the secret society "Skull and Bones," and is constantly calling for more war. He has repeatedly promised to fight the war on terror "better than George Bush" if elected. As explained further in Part VI, the foreign war on terror is really just a cover for the war for the world's rapidly dwindling oil supplies.

Kerry explicitly endorses a highly militaristic foreign policy on page 40 of his biography, *A Call to Service,* where he advocates for the "the tough-minded strategy of international engagement and leadership forged by Wilson and Roosevelt in the two world wars and championed by Truman and Kennedy in the Cold War."[213]

What many Americans fail to realize is that in order for America to be a military power abroad, it must be a fascist police state at home. Jimmy Carter's National Security Advisor Zbigniew Brezinski explains why on page 35 of his 1997 book, *The Grand Chessboard*:

> It is also a fact that America is too democratic at home to be autocratic abroad. This limits the use of America's power, especially its capacity for military intimidation. Never before has a populist democracy attained international supremacy. But the pursuit of power is not a goal that commands popular passion, except in conditions of a sudden threat or challenge to the public's sense of domestic well-being. The economic self-denial (that is, defense spending) and the human sacrifice (casualties, even among professional soldiers) required in the effort are uncongenial to

democratic instincts. Democracy is inimical to imperial mobilization.

Brezinski's philosophy is as fundamental to the Kerry/Edwards neo-liberal camp as Paul Wolfowitz's philosophy is to the Bush/Cheney neo-conservative cabal. Consequently, when Kerry says he'll fight the war on terror "better than George Bush," he is implicitly stating he'll clamp down on dissent even more than George Bush has. The average voter may not make this connection, but I guarantee you the corporate and financial interests backing Kerry sure do.

Kerry also voted for the fascist Patriot Act, was a key figure in the passage of the feudalist NAFTA, and voted yes for the imperialist war in Iraq by unconstitutionally ceding the power to declare war from Congress to the President. When he did, he must have realized that he was helping to create a situation whereby he would have the power to single-handedly declare war should he be elected President.

Despite Kerry's crowd-pleasing and frequently repeated promise to crack down on "Benedict-Arnold CEOs," his real constituency is the same as that of George Bush: wealthy corporations. In an August 2, 2004, interview with *Business Week*, Kerry stated:

> I am going to bring corporate America to the table... to say: How do we make you more competitive? How do we get out of your way? Research-and-development tax credits? I'd make them permanent and larger. Manufacturing tax credits? That's a smart way to help... I am 100 percent in favor of companies going abroad to do business.[214]

Kerry's line about companies going abroad is pretty easy to translate: he's in favor of more outsourcing.

As far as making corporate America more competitive: this can only be achieved if corporate America is given unfettered access to the world's oil supplies. This will only be achieved through massive military intervention, which will almost certainly require a military draft. One would be hard-pressed to think of somebody better suited to call for a draft than an ex-war hero turned anti-war protestor like Kerry. After all, if a President with Kerry's background calls for the draft, it must be necessary, right?

Kerry's running mate, John Edwards, might even be worse. At the 2004 Democratic Convention, he gave a speech that could have come straight from the mouths of George Bush or Dick Cheney. In his speech, Edwards stated, "We must be one America, strong and united for another very important reason. Because we are at war." Translation: dissent will not be allowed under a Kerry/Edwards regime. Edwards went on to say:

We will strengthen and modernize our military, we will double our Special Forces, we will invest in the new equipment and technologies so that our military remains the best-equipped and best-prepared in the world. This will make our military stronger; it will make sure that we can defeat any enemy in this new world.[215]

This paragraph must have sounded like music to the ears of the defense industry.

Jimmy Carter's Convention speech was no less disturbing. He stated, "We need John Kerry to restore life to the Global War on Terrorism." As Sonali Kolhatkar and James Ingalls noted in a recent article for *ZNet.org*, "If the war on terrorism needed any more life than Bush gave it in Iraq and Afghanistan, the world is in for a nightmare."[216]

If the speeches given inside the Convention were disturbing, the scenes outside the convention were downright terrifying. Citizens who arrived to protest the war in Iraq were herded into a "free speech zone." These free speech zones are particularly horrifying for two reasons:

1. By designating a certain area a "free speech zone," the government camp has, by default, designated everywhere else a "non-free speech zone." This is a patent and overt violation of the Constitution. The first amendment to the Constitution designates the entire United States a free speech zone, not specific areas chosen by the government.

2. According to eye-witness reports, the zone resembled a scene from some post-apocalyptic movie: It was surrounded on all sides by concrete blocks and steel fencing, with razor wire lining the perimeter. A giant black net covered the entire space.[217] The federal judge who heard a challenge to the demonstration zone by protest groups on July 22nd stated in open court:

> I, at first, thought before taking the view [of the site] that the characterizations of the space as being like an internment camp were litigation hyperbole. I now believe that it's an understatement. One cannot conceive of what other elements you would put in place to make a space more of an affront to the idea of free expression.

Despite that, the judge denied the groups' challenge to the conditions.[218] One wonders what it would have taken for the judge to rule against the conditions.

Should Kerry win the election, it will likely be because the Bush cabal has so botched the war in Iraq. Once in office, however, Kerry will be free to pursue the same corporatist and militaristic agenda pursued by Bush except he will be unhampered by all the political (body) baggage accumulated by his predecessor. In other words, he really will be able to do it "better than Bush."

Finally, keep in mind that even if we're Bush-free come January 2005, Jeb is waiting in the wings for 2008 and 2012. The only alternative to Jeb will likely be Arnold.

Basically we're screwed either way.

Part VII. Peak Oil and Global War

"Once global oil peaks, and we need to start pumping Saddam's oil, I expect Americans to invade and occupy Iraq. Moreover, profits will flow to friends of George Bush – not some wild-eyed, gun-waving crackpot like Saddam. Obviously, once oil production peaks in a couple of years, the public will throw their total support behind an invasion of Iraq."
-Jay Hanson in 1997

"I know not with what weapons World War III will be fought, but World War IV will be fought with sticks and stones."
-Albert Einstein

"Life without cheap oil is going to be just like life during the Bible. Well, at least the bad parts of the Bible."
-Matt Savinar

"For the few who would like some faint idea of what is ahead, I suggest reading about the savagery which accompanied and followed the fall of the Western Roman Empire (ca. 450-650). When reading, remember: we now have nukes."
-Unknown

"After the Americans destroyed our village and killed many of us, we also lost our houses and have nothing to eat. However, we would have endured these miseries and even accepted them, if the Americans had not sentenced us all to death. When I saw my deformed grandson, I realized that my hopes of the future have vanished for good, different from the hopelessness of the Russian barbarism, even though at that time I lost my older son Shafiqullah. This time, however, I know we are part of the invisible genocide brought on us by America, a silent death from which I know we will not escape."
-Jooma Khan of Afghanistan, March 2003. (Speaking about the effects of Depleted Uranium munitions on Afghan children.)

81. What is the government doing to solve this problem?

The government has two solutions for this problem: go to war to get oil and kill anybody who gets in the way.

A. Go to war to get oil:

Our government's first solution to the coming oil shortages was summarized in April 2001, when a report commissioned by Dick Cheney was released. The report made the following points:

1. "The American people continue to demand plentiful and cheap energy without sacrifice or inconvenience."

2. The possibility of disruptions in the US oil supply are greater than they have been in two decades.

3. The US is running out of energy and can expect more "Californias."

4. The US needs to put energy at the heart of its foreign policy.[219]

The need for military intervention to secure fuel supplies was later echoed by Jeffery Record, a former staff member of the Senate Armed Services Committee, in an article for the Army War College Journal. In December 2002, the *Sydney Morning Herald* summarized Record's article as follows:

> Record argues for the legitimacy of "shooting in the Persian Gulf on behalf of lower gas prices." He also advocates the acceptability of presidential subterfuge in the promotion of a conflict and explicitly urges painting over the US's actual reasons for warfare with a noble, high-minded veneer, seeing such as a necessity for mobilizing public support for a conflict.[220]

As explained previously, PNAC has also advocated the US pursue control of the world's energy supplies through a hyper-militarized foreign policy. The Democratic foreign policy plan isn't that much different except that it might not be as incredibly brazen as PNAC's plan.

In short, our leaders have decided to make a last-ditch grab for what little cheap oil is available by stealing it from the nations that have it. With control over the world's dwindling supplies of cheap oil, they will have the ability to choose who lives and who dies.

B. Forced depopulation of resource rich areas:

Our government's other solution to the coming oil shortages is forced depopulation (genocide) of oil-rich areas. This policy is nothing new and not unique to any one political persuasion.

In 1968, environmentalist Dr. Paul Ehrlich published *The Population Bomb*, which sold about 20 million copies and exerted a huge influence upon policy makers. According to Ehrlich:

> Our position requires that we take immediate action at home and promote effective action worldwide. We must have population control at home, hopefully through a system of incentives and penalties, but by compulsion if voluntary methods fail. We can no longer afford merely to treat the symptoms of the cancer of population growth; the cancer itself must be cut out.[221]

Dr. Ehrlich goes on to say that compulsory birth control could be imposed by governments via the addition of "temporary sterilants to water supplies or staple food."[222]

A few years later, the brutal methods advocated by Ehrlich were officially discussed in the highest levels of our government. In December 1974, the US National Security Council completed a classified 200-page study, "National Security Study Memorandum 200: Implications of Worldwide Population Growth for US Security and Overseas Interests (NSSM)." The study explained that the US needed to control populations in third world countries in order to maintain access to certain resources:

> The location of known reserves of higher-grade ores of most minerals favors increasing dependence of all industrialized regions on imports from less-developed countries. The US economy will require large and increasing amounts of minerals from abroad, especially from less-developed countries. That fact gives the US enhanced interest in the political, economic, and social stability of the supplying countries. Wherever a lessening of population pressures through reduced birth rates can increase the prospects for such stability, population policy becomes relevant to resource supplies and to the economic interests of the United States.[223]

In 1988 the Pentagon released a report entitled "Global Demographic Trends to the Year 2010: Implications for US Security," which advocated that "population planning be given the status of weapons development."[224]

A government report that was reprinted in the Summer 1991 edition of *Foreign Affairs* as "Population Change and National Security," warned that current population trends:

1. Raise serious concerns about "the international political order and the balance of world power;"

2. Could create an "international environment even more menacing to the security prospects of the Western alliance than was the Cold War for the past generation."[225]

When our leaders frame population growth as this extreme of a national security threat, why should it come as much of a surprise that their favored solution is genocide?

Genocide is a strong word, but it is the only term that accurately describes what has happened in Iraq since 1990:

From 1990-2003, the US ensured that heavy economic sanctions were levied against Iraq, ostensibly to deal with Saddam Hussein. While the sanctions had little effect on Saddam, they did have an effect on the Iraqi population, particularly its young children:

1. In 1980, approximately 43,000 Iraqi children under the age of five died. By 1990, this number had dropped to 35,000 per year. In 1991, the year after the sanctions were imposed, the number jumped to 92,000 per year. By the year 2000, that number had jumped to 104,000 per year.[226]

2. Between 1991 and 1998, half a million Iraqi children under the age of 5 died as a result of the sanctions.[227] If the years 1999-2003 are factored in, the number would likely be closer to 1 million. As staggering as that number is, its true magnitude is recognized only when you consider that Iraq's total population is only 24 million. In comparison, the US has a population of about 280 million, or twelve times that of Iraq. A million dead children in Iraq is mathematically equivalent to 12 million dead children in the US.

The death of this many Iraqi children was no accident. Professor Thomas Nagy of George Washington University has completed a study entitled, "The Secret Behind the Sanctions," in which he cites several declassified Defense Information Agency documents now available on the Internet that definitively prove the US intended to kill Iraqi civilians. According to Nagy, these documents clearly demonstrate the US consciously intended to destroy the

Iraqi civilian infrastructure, and in particular, its ability to deliver fresh water.[228]

Since invading and occupying the place, the US hasn't done much to remedy the damage it wrought upon Iraq's civilian infrastructure during the 1990's. In July 2004, the White House released a report explaining that the US government has spent only two percent of an $18.4 billion aid package that Congress approved in October 2003.

As *The Washington Post* reported, "virtually nothing from the package has been spent on construction, health care, sanitation and water projects. More money has been spent on administration than all projects related to education, human rights, democracy and governance."[229]

Failing to spend available money on things like health care and education is exactly what an occupying nation would do if its goal were to depopulate the place.

That a program of depopulation is in place in Iraq is further evidenced when one considers the effects that Depleted Uranium (DU) munitions have had on the Iraqi population. DU is a by-product of nuclear waste. When placed in the tip of a projectile, the projectile acquires armor-piercing capability. When a tank is covered with DU armor, the tank becomes impervious to enemy rounds. When a DU round explodes, it "aerosols," spreading nuclear waste into the air and ground.

During the first Gulf War, the US dropped so much DU on Iraq that in Basra, cancer rates have since jumped by 1,000 percent while infertility rates have doubled.[230] In some cases, the radiation was so bad that 67 percent of American Gulf War veterans ended up having babies with serious birth defects.[231]

In 2003, we dropped so much DU on Baghdad that radiation levels rose to 1,000 times normal.[232] According to the former chief of India's Navy, the total amount of radiation in Iraq in 2003 is equivalent to the amount that would be produced by 250,000 Nagasaki Atom Bombs.[233] DU has a half-life of 4.5 billion years. Essentially, we have eliminated the Iraqi population (and many of our own troops) from the healthy human gene pool.

C. Conclusion

If you have trouble believing the US government is capable of implementing such a heinous plan, consider the following quote from George Kennan, former head of the US State Department Policy Planning:

We have about 60 percent of the world's wealth but only 6.3 percent of its population. In this situation we cannot fail to be the object of envy and resentment. Our real task in the coming period is to devise a pattern of relationships which will permit us to maintain this position of disparity. We need not deceive ourselves that we can afford today the luxury of altruism and world benefaction. We should cease to talk about such vague and unreal objectives as human rights, the raising of living standards and democratization. The day is not far off when we are going to have to deal in straight power concepts. The less we are then hampered by idealistic slogans, the better.[234]

Don't think the sentiments expressed by Kennan are isolated. If anything, they are the rule, not the exception. The dynamics of our electoral system prevent our leaders from developing any plans to deal with this situation that are not predicated on violence and cold, hard, cost-benefit analysis. The biggest campaign contributors are companies from the transportation, energy, defense, and pharmaceutical industries. Corporate officers from these industries find themselves either elected into office or appointed to significant advisory positions. After years in the corporate sector, these individuals have been conditioned to think solely in terms of assets, liabilities and profits. Access to oil = asset; anybody who gets in the way = liability. To be profitable, you must maximize assets and minimize liabilities. It is as simple as that.

82. So the war in Iraq was about oil? Was Saddam the true target?

The theory that the current war in Iraq is simply "blood for oil" is only partially true.

As you may already know, Iraq sits on top of a lot of oil, 115 billion barrels, to be exact. The only country that has more is Saudi Arabia, with an estimated 263 billion barrels. Of equal importance is the fact Iraq's oil production peak won't occur until after 2010. By that time, Saudi Arabia will almost certainly be in decline, thus leaving Iraq as the world's sole swing producer. Since modern, conventional warfare is entirely oil-powered, whoever controls Iraqi oil after the year 2010 will possess a huge geopolitical advantage over the rest of the world.

However, while the desire of the US to control Iraqi oil shouldn't be underestimated, it was not the sole or even primary motivation for the invasion.

The true target of the invasion was the European economy, not Saddam Hussein. The true weapon of mass destruction was the euro, not anthrax.

To understand why, you must understand how the "petrodollar" works. All oil transactions have been priced in dollars since World War II. For the last 60 years, anytime anybody wants to buy oil, anywhere in the world, they have to pay in US dollars. This has created an artificial demand for the dollar, thus raising its value and giving the US control over global oil sales.[235]

This system evolved because the US was originally the world's number one oil producer and oil exporter. Since the majority of the world's oil was coming from under US ground, it simply made sense to price oil transactions in dollars.

Although the US ceased being the world's number one producer/exporter over a generation ago, it has refused to let go of the petrodollar. There is good reason: Because we no longer manufacture anything, we have nothing to sell to the rest of the world. The only thing the US has that it can legitimately sell is oil! Consequently, the US economy is entirely dependant on the petrodollar.

Some of you may be scratching your head, "But I thought most of the world's oil comes from other parts of the world? How can the US be selling oil if it is extracted and sold by other countries?" **Remember, all oil transactions are priced in dollars**. Thus, even if the oil comes from another part of the world, because it has to be paid for in dollars, it's as though the US is selling it.

Consequently, any country that attempts to price oil transactions in the euro poses an extremely ominous financial threat to the US economy and will be dealt with accordingly.[236]

In November 2000, France persuaded Saddam to switch from the dollar to the euro as the currency for its oil transactions. This caused the euro to gain considerably against the dollar.[237] It's probably no coincidence that the presidential candidate more likely to go after Saddam was installed in office shortly thereafter.

Moral concerns aside, the Bush Administration's complete determination to invade Iraq makes a lot more sense when seen in the context of these facts. Many people feel the Bush-led invasion of Iraq has undermined our national and economic security. What they fail to realize is that the invasion temporarily dissuaded OPEC from wholeheartedly embracing the euro. Had Bush not invaded Iraq, OPEC would likely have embraced the euro, the dollar would have collapsed, and the US would have lost whatever national and economic security it has left.

83. Iraq, Afghanistan... Who else is on the hit list?

Any country that meets at least two of the following criteria:

1. Sits atop significant oil reserves, like Iraq does.

2. Has attempted or threatened to price oil transactions in euros, like Iraq did.

3. Is accused of "harboring terrorists," "developing WMD," or having connections to 9/11, like Iraq was.

4. Is discussed in documents released from Dick Cheney's secretive "Energy Task Force," as Iraq is. If a country's oil fields are mapped out and assigned to certain companies in documents released from the Task Force, as Iraq's are, an invasion is all but guaranteed.

5. Is mentioned by PNAC in their manifesto, "Rebuilding America's Defenses."

84. Is the US going to invade Iran?

Yes, for the following reasons:

1. Iran has the world's second- or third-largest oil reserves, depending on whose statistics you believe. According to the Iranian Oil Ministry, the country has 132 billion barrels. According to the OPEC Website, Iran has 99 billion barrels. Either way, Iran is sitting on a lot of oil, just like Iraq.

2. Iran now accepts euros for its oil exports. In addition, it has announced that, in 2005, it will launch an oil-trading market for Middle East and OPEC producers which will likely be euro-based.

3. The Iranian government has admitted it is pursuing nuclear technologies.

4. In "Rebuilding America's Defenses," PNAC explains that Iran may end up being a bigger threat to the US than Iraq.

The American public is already being propagandized to accept the invasion of Iran as recent news reports have accused Iran of having had a hand in the 9/11 attacks.

The Iranian government is fully aware that it has been targeted for invasion. Iran's supreme leader, Ayatollah Ali Khamenei, didn't make anybody read between the lines when he recently warned the US that, "If the enemy attacks our scientific, natural, human or technological interests, the Iranian people will cut off its hand without hesitation and place in danger the interests of the aggressor everywhere in the world."[238]

On a similar note, Ali Samsam Bakhtiari, the vice-president of the national Iranian oil company has stated, "The crisis is very, very near. World War III has started. It has already affected every single citizen of the Middle East. Soon it will spill over to affect every single citizen of the world."[239]

I often get emails from people asking me, "When will the world descend into a Mad-Max-style existence?" I generally respond by telling them their question is akin to a person in 1863 Maine asking when the Civil War was going to start. As Mr. Bakhtiari pointed out, much of the world already has descended into a Mad-Max-style existence. Just because your neighborhood isn't yet overrun by marauding gangs doesn't mean the descent hasn't already started.

85. Is the US going to invade Saudi Arabia?

Yes, for the following reasons:

1. Saudi Arabia sits on top of large oil reserves, like Iraq. Its oil fields were mapped out by Dick Cheney's Task Force, just like Iraq's oil fields.

2. A plan to invade Saudi Arabia and seize its oil fields has been on the books since the oil shocks of the 1970s.

   In 1973, the Nixon administration developed a plan to invade Saudi Arabia and seize its oil fields. The plan, known as "UK Eyes Alpha," was declassified in December 2003. It explains that the US could secure its oil supply by seizing the oil fields in Saudi Arabia and the United Arab Emirates.[240]

3. The desire to invade Saudi Arabia has not lessened over the past 30 years:

   In their November 2000 report, PNAC advocates for the establishment of US military bases in Saudi Arabia.

   In July 2002, the RAND Corporation's Laurent Murawiec gave a PowerPoint presentation to the Defense Policy Board entitled "Taking Saudi Out Of Arabia." In his presentation, Murawiec advocated the US invade Mecca and Medina, confiscate Saudi Arabian financial assets, and seize its oil fields.[241]

In 2003, former US diplomat James Akins was quoted in the liberal leaning, albeit largely reliable, *Mother Jones Magazine* as saying, "It'll be easier once we have Iraq. Kuwait, we already have. Qatar and Bahrain, too. So it's only Saudi Arabia we're talking about, and the United Arab Emirates falls into place."[242]

4. As explained in Part II, the House of Saud is extremely unstable. If it falls, chaos will ensue in Saudi Arabia and oil prices will quickly reach cataclysmic levels. Both the US government and corporate America are acutely aware of this possibility. For instance, according to Robert E. Ebel, director of the energy program at the highly influential Washington D.C. think tank Center for Strategic and International Studies, "If something happens in Saudi Arabia . . . if the ruling family is ousted, if they decide to shut off the oil supply, we have to go in."[243]

86.   Is the US going to invade Venezuela?

Possibly, for the following reasons:

1. Venezuela is the fifth-biggest oil producer in the world, and is the third-largest supplier to the US, exporting oil to this country at the rate of 1.5 million barrels per day. That's about three-quarters as much oil as Saudi Arabia exports to the US per day.

2. Venezuela's President, Hugo Chavez, is supporting an OPEC move to switch from the US dollar to the euro.

   What's worse, at least as far as the US is concerned, is that he accepts oil-for-service swaps with poor countries, such as Cuba, that don't have sufficient US dollar reserves with which to buy the oil it desperately needs, effectively avoiding the use of the petrodollar.[244]

In March 2004, President Hugo Chavez vowed on national television to freeze oil exports to the United States and wage a "100-year war" if Washington ever tried to invade Venezuela. If Saudi Arabia's oil production plummets at any point, an invasion of Venezuela will become a very real possibility.

87.   Is the US going to invade West Africa?

Yes, for the following reasons:

1. West Africa currently supplies the US with 15 percent of its oil imports. This number is expected to jump to 25 percent by 2015.[245]

2. While Africa's oil reserves are not as plentiful as those in the Middle East, they have a much shorter "field-to-tank" time. Whereas it takes about 6 weeks for oil from the Middle East to get to the US, oil from West Africa only takes about 2 weeks.[246]

3. Islamic fundamentalism is well rooted in West Africa, particularly in Nigeria, which is the area's number-one oil producer.

4. The US has already established a string of military bases in and around West Africa in order to protect its oil interests.[247]

88.   Is the US going to invade Canada?

Quite possibly, but not for a while.

The US gets 87 percent of its natural gas imports from Canada, who is required by NAFTA to sell 60 percent of its natural gas to the US. When Canada begins to experience the energy crisis, they may seek to change the terms of that law. The US is unlikely to allow them to do so.

Most Americans are shocked to find out the US imports a greater amount of oil from Canada than from Saudi Arabia. As of 2002, the US imports 1.8 million barrels per day from Saudi Arabia, and 2.1 million barrels per day from Canada.

The reason Cheney and Rumsfeld have stated the "war on terror" will last 40 years is because by 2040-2050, the bulk of the world's recoverable oil will be in Canada. By the year 2050, over half of the world's oil supply will come from so-called "non-conventional" oil. Most of this oil is located in Canada, with some located in Venezuela. Consequently, whoever controls Canada in 2050 will control the world, at least what's left of it.

The initial signs of a potential US-Canadian conflict are already appearing. In August 2003, the US initially blamed the New York blackout on Canada. In July 2004, *Petroleum News* reported low-level tensions between Canada and the United States over ownership of possible offshore oil and gas deposits in the Beaufort Sea have intensified from "simmer to boil."[248] Less than one month later, the *New York Times* reported that Canada has initiated large-scale military operations in the previously isolated and forgotten Arctic regions. The reason? To reinforce its claim to oil and gas reserves the US and EU have both claimed as their own.[249]

If a war with Canada seems impossible, consider the following for a bit of historical perspective: at the turn of the 20th century, Germany was considered one of the most socially advanced, benevolent, and civilized nations in the world. Thirty years later, it had degenerated into one of the most barbaric states in history. The German people, who were starving during the 1920s, embraced the barbarism because they had been promised it would put food on their tables. The human mind, regardless of whether it resides in a German body in 1933 or an American body in 2023, can get pretty warped when food gets scarce.

Think about it this way: if your family was starving, and you saw that your neighbor had food, would you consider taking it by force? Most people would. Remember that, as explained in Part II, oil production = food production. If Americans can't get food (oil), we're going to take it from whoever has it.

89.   Is the US going to invade France?

France doesn't have any oil reserves worth controlling, so probably not. However, as explained previously, the war in Iraq is the first battle in the US-led war against the EU. This should not come as a surprise, as several high-level officials in the Bush Administration have been publicly pushing a plan to force nations to "choose between Paris and Washington."[250] The US wants to make sure that Saudi Arabia, who exports more oil to the EU than it does to the US, chooses Washington.

Not surprisingly, some of the officials who wrote the above-cited report are members of PNAC.

90.   What's going to happen when recently industrialized China decides it needs what little cheap oil is left as bad as the US does?

World War III.

A war with China may seem absurdly suicidal but that's exactly what PNAC is not so subtly advocating when they say the US should "spur democratization" in China.

A silent energy war between the US and China has already broken out on a full scale. Figuratively speaking, the first shots in this war were fired during the lead-up to the US-led invasion of Iraq. China vehemently opposed the invasion, not because of any humanitarian concerns, but because the China National Petroleum Company had long sought to secure major oil supplies from Iraq.

According to China's Ministry of Commerce, its oil imports rose by 30 percent last year. Despite the extra oil, as well as millions of tons in increased domestic coal production, China already has begun to suffer from severe energy shortages and widespread blackouts.

In recent months, China and Saudi Arabia have signed several oil and natural gas contracts. These agreements make the US nervous as it depends on two key allies in the region, Saudi Arabia and Egypt, for control of the Middle East. Some signs suggest that Saudi Arabia and China are developing a weapons-for-oil deal.

As if that wasn't bad enough, the Chinese government gave the euro its much-coveted seal of approval in July, announcing that it would switch part of its vast dollar reserves into the world's emerging "reserve currency."[251]

Shortly after China decided to back the euro, the US Navy began conducting "Operation Summer Pulse '04 off the China coast near Taiwan. In total, the Navy deployed seven of its 12 carrier strike groups.

As Chalmers Johnson points out in a *Los Angeles Times* article dated July 15, 2004, such a huge deployment is extremely ominous, "Typically, if a crisis erupts, the Navy will deploy one or two carrier groups to a particular region. During a war it might deploy three or four, as it did for both wars in Iraq. Seven at once is absolutely unprecedented."[252] The US military is clearly preparing for something big.

Some people hope that increased economic interdependence between China and the US will prevent the two nations from going to war. After all, several of the big US automobile companies want to sell cars to the rapidly growing Chinese market. These companies would stand to lose profits if the US and China went to war, right? To prevent this, they would likely lobby the US to not go to war with China, in which case a tense peace could be maintained, right?

Don't get your hopes up. In the 1920s and 1930s, US companies were heavily invested in Nazi Germany and that didn't stop World War II. In fact, it may have created conditions that encouraged the war. Rather than divesting from the Nazi regime, these companies made tons of money selling goods and resources to both sides of the war.

As the following excerpt from a report printed by the United States Senate Committee on the Judiciary in 1974 explains:

GM's plants in Germany built thousands of bomber and jet fighter propulsion systems for the Luftwaffe at the same time that its American plants produced aircraft engines for the US Army Air Corps . . .

Ford was also active in Nazi Germany's prewar preparations. In 1938, for instance, it opened a truck assembly plant in Berlin whose "real purpose," according to US Army Intelligence, was producing "troop transport-type" vehicles for the Wehrmacht.

The outbreak of war in September 1939 resulted inevitably in the full conversion by GM and Ford of their Axis plants to the production of military aircraft and trucks. On the ground, GM and Ford subsidiaries built nearly 90 percent of the armored "mule" 3-ton half-trucks and more than 70 percent of the Reich's medium- and heavy-duty trucks.[253]

It's not just the automobile manufacturers who will make out like bandits by selling military hardware to all sides of the coming resource wars. Historically, large technology companies have done as well, if not better, than vehicle and weapons manufacturers during wartime. IBM, for instance, profited handsomely from the Holocaust. Beginning in 1933, they developed a system of punch cards and automated technology that allowed Hitler to systematically locate Jewish families, round them up, and kill them. If you're interested in verifying this for yourself, get a copy of the book *IBM and the Holocaust* by Edwin Black. Excerpts and extensive supporting documentation are available online, for free, at http://www.ibmandtheholocaust.com.

Black's book has been reviewed by *The Los Angeles Times, Newsweek, The Chicago Tribune, The Nation, The Miami Herald,* and dozens of other mainstream, reputable news sources. All agree that his research is absolutely impeccable and beyond reproach.

Remember, just as in the 1930s and 1940s, we are dealing with "multinational" companies. Just because you think of them as "US companies" does not mean they think of themselves as such. Their allegiance is to the bottom line, not to a particular country or political ideology, let alone to any sense of human decency or morality. Legally, they are required to do what is in the best interests of the company, not what is in the best interests of the country they are typically associated with. Furthermore, they don't care who "wins" the war. They can make as much money selling a bullet to China to kill a US serviceman as they can make selling a bullet to the US to kill a Chinese serviceman. If they can figure out how to sell weapons, vehicles, and technology to both sides of the war, they can double their profits and still be in the good graces of which ever country comes out on top once the smoke clears.

The reason I cite so much historical precedent in a book about the future is to get you to understand a very chilling but absolutely unavoidable fact about the coming oil wars: our leaders are not going to attempt to deal with the world's resource constraints through peaceable methods because these methods don't offer much in the way of profit. In contrast, a massive petro-apocalypse promises to be hugely profitable for our leaders.

Take Tom Ridge for example. According to disclosure forms known as the Executive Branch Personnel-Public Financial Disclosure Report, Ridge has investments in Raytheon (maker of the Tomahawk missile), General Electric (maker of nuclear weapons) and several other companies the US government has given homeland security related contracts to.[254]

Think about that for a moment. **The man charged with securing the homeland is invested in nuclear bombs.** One wonders what future anthropologists will say when they dig this factoid up. In fairness to Ridge, he is only one of hundreds high ranking elected and appointed officials heavily invested in the arms industry. George Bush 41, for instance, went to work as a consultant for the Carlyle Group (a major defenses contractor) after his term in the White House.

Six billion dead people will equal a lot of "shareholder value" for the companies supplying the hardware and resources necessary for the killing. It will also raise the salaries of the employees, particularly the salaries of the (former) elected and appointed officials who are either invested in these companies or who will go work for these companies after they initiate the killing spree during their time in public office.

Yes, that is really how the racket works: our leaders vote for war during their time in public office, knowing that once they return to the private sector they will have the opportunity to work for the same industries and companies that are profiting from the killing spree they voted for while in office. Naturally, our leaders want these industries and companies to do well as they will eventually be working for them. The more wars our leaders initiate while in office, the more weapons these companies will sell. As these companies sell more weapons, they raise the salaries of their employees, including those employees who just got done with a stint as President, in Congress, as a congressional staffer, or as a high-level government appointee.

A large-scale, multi-decade proxy war between the US and China will likely be the most profitable time in history for many of these companies and the former elected officials who work for them. It promises to be far more profitable than even World War II simply because there are so many more people to kill than there were 60 years ago. The equation is simple: more weapons + more vehicles + more computers + more body bags = more profits, which is what the law requires corporate officers to pursue above all else. In short, the law precludes us from avoiding World Oil War IV.

91. Well, at least we don't have to worry about Russia anymore. After 9/11, they said they would support us. They're our friends now, right?

If Russia is considered a "friend," the US doesn't need any enemies. Russian President Vladimir Putin has been building up Russia's nuclear capability since 1999 because he (justifiably) fears the US is trying to muscle in on the oil reserves located in the Caspian Sea.

Russia flexed its muscles after the April 2002 summit to try to settle the Caspian Sea issues. Within hours of leaving the summit, President Putin launched the largest naval exercises in the Caspian Sea's history[255]

In February 2004, Russia's nuclear forces began preparing their largest maneuvers in two decades, including a massive simulation of an all-out nuclear war.[256]

In August 2004, the Russian government announced a plan to double the nation's GDP by 2012.[257] For the plan to succeed, Russia will need unfettered control of the world's oil supplies, which will require tremendous investments in new armaments. Not surprisingly, President Putin recently announced a plan to increase Russia's military procurement budget by 40 percent during 2005.[258]

Russia has also signaled that it intends to price its oil and gas exports in euros. The move is part of a deal between President Putin and German Chancellor Gerhard Schroeder aimed at challenging American global dominance.[259]

Remember that oil production inside the Soviet Union reached its peak in 1987. This put the already unstable Soviet economy in a very precarious position. Hoping to capitalize on this turn of events, the US persuaded Saudi Arabia to flood the market with cheap oil. The resulting drop in price was a primary, albeit rarely discussed, factor in the collapse of the Soviet Union.[260]

Many high-level officials in the Russian government remember the "good ole days" when they were at the helm of one of the world's two superpowers. They remember well that US foreign policy played a hand in the Soviet collapse. They would like nothing more than to give the US a dose of its own medicine. By uniting with countries such as France and China (see next question), they might get their chance.

92. Is it possible that China and Russia might get together and gang up on the US?

Yes.

Concerned about the US pursuit of "Star Wars," China and Russia met in January 2001 (eight months before 9/11) to proclaim their friendship amidst serious concerns about hyper-militarized US foreign policy.[261] Those concerns have deepened considerably in the past three years, as the US has become increasingly aggressive, China has become increasingly industrialized, and oil has become increasingly scarce.

A Sino-Russian alliance is now more likely than ever because, by the fall of 2004, the two nations were close to completing agreements on the construction of a major oil pipeline in addition to other energy deals.

If a war between China and the US will make tons of money for big business, a war between China-Russia, and the US, is almost too good to be true, at least as far as the defense industry is concerned.

93. You forgot about North Korea.

Oh yeah. Them, too. PNAC has targeted them for regime change, they either have nukes or are close to having them, and Kim Jong made the same mistake Saddam made: he embraced the euro in January 2002.[262]

94. Gosh, this sounds worse than World War II.

Speaking of which, there is also Japan, which is second only to the US in terms of yearly oil consumption. When the time comes, Japan is more than willing and able to defend its oil interests. Few people realize that Japan has the world's third-largest military. While Japan is only allowed to spend 1 percent of its GNP on defense, it has a huge GNP, along with the world's most advanced technological base.

It also has a highly militaristic history and culture to go with huge stockpiles of plutonium. If Japan wants to, it can construct nuclear weapons within one week.

Japan is already engaged in an undeclared war with its old rival China over access to Russian oil reserves. Japan needs the oil because it has none of its own. China needs the oil to feed its rapidly growing economy.

141

In February 2004, Japan signed an agreement with "axis of evil" member Iran to allow Japan to develop a major oil field in southwestern Iran. As *CNN International* reported, the agreement will provide a "key source of oil for resource-poor Japan, which is also pursuing similar arrangements in Russia and other countries."[263] The agreement drew considerable concern from the US which fears Iran will use the agreement to pursue nuclear weapons development.

Lest we forget history, the US entered World War II after Japan attacked Pearl Harbor, in retaliation for Franklin Roosevelt cutting them off from their oil supply. World War II ended when the US dropped atomic bombs on Hiroshima and Nagasaki.

Don't think Japan has forgiven or forgotten this. Nor will its leaders fail to capitalize on the memory of World War II to justify an increasingly aggressive foreign policy.

95. Isn't this going to require a reinstitution of the draft?

Yes.

For several years now, every young man and woman has been earmarked as a soldier for future oil wars. As the following pieces of evidence demonstrate, the march towards reinstitution of the draft has been underway for some time. Unfortunately, few people started paying attention until recently.

In May 2003, the chief of the Selective Service, Lewis Brodsky, gave a presentation to the Department of Defense in which he recommended the military draft be expanded to include all men and women, ages 18-34. Brodsky acknowledged the plan might have to be "marketed."[264]

On September 23, 2003, the Pentagon placed a notice on its Website asking for "men and women in the community who might be willing to serve as members of a local draft board." The Pentagon quickly pulled the announcement from its Website after public outcry.

As of March 2004, the "Universal Military Training and Service Act" is circulating in Congress. If passed, the bill would make military service a requirement for all men and women (including college students) between the age of 18 and 25.[265]

According to its budget for fiscal year 2004, the Selective Service has received 28 million dollars to implement "performance measures," most of which seem aimed at improving its ability to conduct a draft. An annual report providing the results of these measures is due March 31, 2005.[266]

In 2003, the Army issued 40,000 "stop-loss" orders.[267] Stop-loss is also known as "draft-creep" because stop-loss orders prohibit troops who are scheduled to retire from doing so until the military releases them.[268]

You can forget about running to Canada. The government already has that covered: in December 2001, Canada and the US signed the "Smart Border Declaration."[269] The plan is ostensibly aimed at stopping terrorists, but many of its provisions will likely be used to make escaping to Canada more difficult once the draft is reinstituted.

Electing John Kerry won't stop the draft. On December 16, 2003, Kerry all but promised he would reinstitute the draft if elected President. In a speech at Drake University, in Des Moines, Iowa, Kerry said:

> As we internationalize the work in Iraq, we need to add 40,000 troops — the equivalent of two divisions — to the American military in order to meet our responsibilities elsewhere, especially in the urgent global war on terror. In my first 100 days as President, I will move to increase the size of our Armed Forces.[270]

If that's not a call for a draft, I don't know what is.

96. Is the US capable of winning these wars?

No.

According to a study performed by the US Army itself, the US war machine has almost completely broken down in Iraq. The report indicates that US troops have had to scavenge for parts and supplies from abandoned trucks and captured Iraq weapons. Moreover, the powerful armor and advanced technology of the US military has been largely offset by Iraqi guerrilla tactics, easily obtained rocket-propelled grenades, and homemade "Improvised Explosive Devices."[271]

One can only wonder what would have happened had Saddam actually put up a fight. The Chinese with their standing million-man army and the Russians with their vast arsenal of Cold War weapons can be counted on to inflict a lot more damage than the poorly trained and under-equipped Iraqi insurgents have.

Don't think the inability of the US to win these wars will stop us from provoking them. Without unfettered access to the world's dwindling supply of fossil fuels, the debt-ridden US financial system will quickly be reduced to a pile of rubble. Consequently, as far as our corporate and governmental elite are concerned, there is no choice but to fight. Their belief is reinforced by the fact most of them stand to profit from the mess.

97. Are we really crazy enough to fight an all-out nuclear war for oil?

Yes.

Remember that in the 21st century oil production = food production. So the question should be, "How many people, in which nuclear-armed nation, are going to starve before they feel they have nothing to lose by initiating a nuclear war?"

Of course, the nuclear war could very well be accidental. With 30,000 nuclear weapons floating around the world, many in the possession of highly aggressive, unstable, and thoroughly unaccountable individuals, an accident is almost inevitable.

Most people are completely unaware that on January 25, 1995, we came within three minutes of witnessing an accidental nuclear holocaust. As Dr. Helen Caldicott explains in her book *The New Nuclear Danger*, on that day Russian radar stations mistook an American missile carrying a scientific probe for a nuclear weapon. Russian military officers accessed the nuclear launch codes and directed Boris Yeltsin to get his finger on the proverbial red button. Yeltsin was within three minutes of launching a retaliatory attack when the US probe veered off-course.[272]

As a result of its subsequent economic collapse, Russia's missile control systems have completely fallen apart since that day and are now highly vulnerable to both accidents and terrorism.[273] At times, the US systems haven't been much more reliable. As the *New York Times* has reported, part of America's nuclear warning system failed to properly function for almost a year during 1999-2000.[274]

Unfortunately, tensions will only heighten as oil becomes more scarce, thereby increasing the chance of accidental nuclear annihilation.

98. Is there any chance we will resolve this situation without an all-out world war?

As things stand now, absolutely not.

Wars are most often fought over life-giving resources such as food and energy. Humanity is facing the most severe resource shortages in world history. There is no reason to expect we won't address these shortages as we have addressed all previous shortages: with warfare. Since the coming shortages are unprecedented in their severity, we can expect the accompanying wars to be unprecedented in their severity as well.

We may not like to admit it, but human beings can be absolute animals. The truth of this has been long confirmed by anthropologists. For instance, in his book *The Dark Side of Man,* author Michael P. Ghiglieri writes:

> War analyst Stanislav Andreski concluded that the trigger for most wars is hunger, or even 'a mere drop from the customary standard of living.' Anthropologists Carol and Melvin Ember spent six years studying war in the late 1980s among 186 pre-industrial societies. They focused on pre-contact times in hopes of collecting the 'cleanest, least distorted' data. Andreski, it seems, was right. War's most common cause, the Embers found, was fear of deprivation. The victors in the wars they studied almost always took territory, food, and/or other critical resources from their enemies.
>
> This also holds true among modern nations. For instance, after studying recent global conflicts, political scientists Thomas E. Homer-Dixon, Jeffrey H. Boutwell, and George W. Rathjens concluded, "There are significant causal links between scarcities of renewable resources and violence. In short, many wars seem to be a mass, communal robbery of another social group's life-support resources.[275]

It seems that when humans are faced with resource shortages we react exactly as our primate cousins do: by killing each other.

The US Army is as aware of this fact as the anthropologists are. In 1997, Major Ralph Peters wrote in the US Army War College's journal, *Parameters*, that during the 21st century:

> There will be no peace. At any given moment for the rest of our lifetimes, there will be multiple conflicts in mutating forms around the globe. Violent conflict will dominate the headlines, but cultural and economic struggles will be steadier and ultimately more decisive. The de facto role of the US armed forces will be to keep the world safe for our economy and open to our cultural assault. To those ends, we will do a fair amount of killing.[276]

James Woolsley, the former chief of the CIA, is more or less in agreement with Major Peters. In April 2003, Woolsley was quoted by *Toronto Star* columnist Linda Diebel as saying, "This Fourth World War, I think, will last considerably longer than either the First or Second World Wars did for us. Hopefully not the full four decades of the Cold War."[277]

99. I think I'm going to be sick.

I know the feeling.

100. Gosh, this sounds like some type of Mad-Max scenario.

You're not the only who thinks so. According to former Australian Prime Minister Keating, "There is every chance that the American policy will lead us into a Mad-Max world, while the US seeks to cocoon itself behind a screen of national missile defense."[278]

Mr. Keating's assertion notwithstanding, such comparisons tend to be problematic as they trivialize the seriousness of our situation. History, not Hollywood, is likely the best guide for what we should expect. As mentioned previously, any good book on the fall of the Roman Empire should provide you with a reasonable approximation of what the next 5-50 years will be like. Factor in modern-day weaponry, and you can see that we have a real mess on our hands.

Part VIII. Managing the Crash & Coping with the Ramifications

"This is preeminently the time to speak the truth, the whole truth, frankly and boldly. [We should not] shrink from honestly facing conditions in our country today. First of all, let me assert my firm belief that the only thing we have to fear is fear itself – nameless, unreasoning, unjustified terror which paralyzes needed efforts to convert retreat into advance."
-Franklin D. Roosevelt

"Courage is resistance to fear, mastery of fear, not absence of fear."
-Mark Twain

"Real peace in a petroleum-fueled world means rejecting petroleum dependence in all possible ways."
-Jan Lundberg

"In the end, optimism is most useful as a state of mind that fosters constructive action; it is self-delusional to dwell on hopeful images of the future merely to distract ourselves from facing unpleasant truths."
-Richard Heinberg

"In life there are going to be some things that make it hard to smile. But whatever you do, through all the rain and pain, you got to keep your sense of humor."
-Tupac Amaru Shakur

"Everyone carries a part of society on his shoulders; no one is relieved of his share of responsibility by others. And no one can find a safe way out for himself if society is sweeping towards destruction."
-Ludwig von Mises

101. I'm by nature an optimist. This all sounds so pessimistic.

If you think discussing Peak Oil and its likely ramifications are too "pessimistic," ask yourself:

1. Was Winston Churchill being a "pessimist" in 1940 when he told Britain, "I have nothing to offer you but blood, toil, tears, and sweat?"

2. Was Albert Einstein being a "pessimist" in 1939 when he told FDR that Nazi Germany was in the process of developing an atomic bomb?

There is a difference between an "optimist" and a fool. An optimist is somebody who looks at bleak facts and decides to make the best of the situation they can. A fool is somebody who looks at bleak facts and decides to ignore them because they are too upsetting.

102. It sounds like this is a serious problem, but you know, I could get cancer or be run over by a car or any number of other catastrophes. Why should I worry about this, too?

Contracting a life-threatening disease, being run over by a car, or experiencing some similar catastrophe is a possibility that may affect you someday. Oil depletion is a certainty that is already affecting you, and is likely to affect you much, much more in the future. Dismissing it now will ensure it hits you harder in the future.

103. I have work, school, bills, kids, traffic, etc., to deal with. How am I supposed to prepare for the oil crash when I'm barely keeping up with life as is?

Join the club. You're not the only person who has day-to-day problems. If Peak Oil is too much for you to worry about, feel free to ignore the facts and stick your head in the sand. Remember, however, that when you stick your head in the sand, you leave your ass exposed for the world to kick.

104. Should I be getting a gun and hiding in the woods?

No, but we are all going to be forced to become more self-sufficient in the years to come. This means being our own farmers, our own doctors, and yes, our own police officers.

While hiding in the woods is probably not the best survival strategy, you will have to take responsibility for your own safety. As the economy dissolves, governments will eliminate police services for everybody except government officials and the super-rich. Consequently, the comfortable confines of the suburbs will begin to resemble lawless, concrete jungles.

If you're a lifelong city slicker like myself, and are not even sure where to start when it comes to evaluating farmland, you might want to check out George Ure's article, "We Finally Bought the Farm." The article is available at www.independencejournal.com/buyfarm.htm. If you'd like to start looking for some farm land to buy, you might want to check out www.unitedcountry.com.

Keep in mind, however, living on a rural homestead will not be free from peril or challenge by a long shot. When our society collapses, the rural areas may well go first. In that case, small enclaves of homesteaders sitting on stockpiles of food, weapons, and gold will be too tempting a target for the bandit cultures that evolve in post-collapse rural areas.

Speaking of bandit cultures, you can be assured that your in-laws will come looking for food and supplies if you have them stockpiled.

It is worth noting that when Rome collapsed, those living in the rural areas suffered more than the city-dwellers. One of Rome's "solutions" to its collapse was the institution of highly oppressive taxes on rural farmers. In the years leading up to the collapse, taxes grew so high that many farmers simply abandoned their land. The situation deteriorated to such a degree that Rome passed laws forcing farmers' children to work as farmers as well.

105. Do you think people will wake up in time for us to avert, or at least soften the crash?

I hope so, but I'm not counting on it because the effects of oil depletion are a lot like the effects of dehydration. When you get dehydrated, your body doesn't completely exhaust its stores of water. Rather, it simply no longer has the amount it needs to function properly. Signs of dehydration include flu-like symptoms such as lethargy and headache. Amazingly, you can die from dehydration without ever feeling thirsty as disorientation often accompanies the lethargy and feelings of malaise.[279]

Once the disorientation kicks in, you could be on the verge of death and wouldn't even know why. In your disoriented state, any solutions you attempt to implement may make your condition drastically worse. For instance, you might blame your headache on stress and attempt to treat it with an aspirin washed down by a cup of coffee. Your "solution" would temporarily mask one of the symptoms (headache) of dehydration but it would ultimately bring you closer to death as caffeine is a diuretic whose effects are amplified by aspirin!

In regards to Peak Oil, you can already see signs of civilization's "petro-dehydration": the economy is slowing down, infrastructure is crumbling, and most importantly, people are disoriented. Some people blame our current problems on terrorists and environmentalists. Others blame our problems on corporations and politicians. All sorts of solutions have been offered but many of them, such as the "hydrogen economy," will only bring us closer to collapse as they waste more energy than they create. In this regard, such projects can be accurately characterized as "petro-diuretics."

Since we are collectively disoriented it is almost impossible for us to understand the true nature of our problem. In fact, we may be worse off than the dehydrated individual as he or she most likely has some idea of how important water is to their everyday life. In contrast, few citizens of petrochemical civilization have any idea how important oil is to their everyday life.

There is no guarantee we will come out of our collective disorientation long enough to accurately diagnose the real cause of our problem: a fundamentally unsustainable economic model that no longer has access to the constantly increasing quantity of resources it requires.

Nor is there any guarantee we will develop viable solutions to our problems once we do wake up. Even if we do come up with a collective plan of action, implementing that plan will involve so much economic and psychological pain that it is highly unlikely we will ever stay the course.

In short, even if we do manage to wake up from our collective fossil fuel fantasy, there is absolutely no way we are all going to get out of bed.

Most of us would rather die in our sleep.

106. What should governments be doing to address these problems?

Forget about the government helping us deal with Peak Oil.

When I wrote the first edition of this book, I included several pages of "policy recommendations" that, if implemented by governments, might help soften the crash.

I have since decided it is absolutely futile to lobby state and national governments to address this issue. These governments are completely controlled by petrochemical interests such as the transportation, agribusiness, pharmaceutical, oil, and defense industries. All the steps we need to take to effectively cope with Peak Oil are at direct odds with the interests of these industries.

For instance, we need to drive less. That's fairly obvious, but what politician is going to advocate a massive reduction in personal use of the automobile and thereby lose campaign contributions from the auto industry?

He would also lose votes en masse as at least 25% of all US jobs are related to the motor-vehicle-industry. A far greater percentage of US jobs are dependant on motor vehicles. Who's going to vote for somebody who promises to pass legislature that might result in the loss of jobs?

Likewise, we need to reduce our population. Again, a fairly obvious idea, but what politician is going to advocate population reduction and thereby lose campaign contributions from the fast-food industry who needs young eaters to support shareholder value or from senior citizens who need young workers to support old-age entitlement programs?

Look at the steps governments have already taken to address this. All of them are aimed at controlling the threat you pose, not at helping you.

The only solutions state and national governments are going to come up with will be variations of the following:

1. Military intervention to secure more oil.

2. Fascist-style legislation to control the citizenry.

3. More wasteful mega-projects like the "Hydrogen Economy" and mission to Mars which are designed solely for the benefit of petrochemical industries that might see their profit margins hurt by oil depletion.

I'm sorry, folks, but we are totally on our own here. The only people you should be lobbying are your neighbors and local community leaders.

107. What are some steps that I can personally take in the next few days to begin addressing this situation?

The following list is by no means exhaustive. These are just some simple steps you can begin taking immediately.

1.   Continue to educate yourself about Peak Oil and its ramifications. A simple Google search will bring back many excellent sites.

2.   Educate others. If you're not sure how to go about doing so, consider lending them this book or emailing them a link to lifeaftertheoilcrash.net. The documentary *End of Suburbia* is also a good way to introduce others to Peak Oil. The film is available both from my site, lifeaftertheoilcrash.net, as well as endofsuburbia.com and postcarbon.org.

3.   Seek out like-minded folks. There are Peak Oil groups forming all over the country. I have a list of these groups available on my site, lifeaftertheoilcrash.net/groups. If there is no group in your area, you may have to step up and form one yourself.

4.   Perform Google searches for Peak Oil whenever you get the chance. As more people search for "Peak Oil," the folks at Google will take notice. This may result in increased mainstream media coverage.

5.   Adopt a vegetarian/vegan diet, or at least reduce your meat consumption as much as you can. Meat is an extraordinarily energy-intensive form of food, and is likely to become quite expensive in the years to come. You may as well begin weaning yourself off of it now.

6.   Start using your bicycle or public transportation instead of your car, whenever possible. If your community has a car cooperative, join it. If your community doesn't have a car cooperative, start one.

7.   Limit your purchase of consumer items as much as you can.

8.   Reduce your use of electricity as much as possible. Consider investing in items such as solar-powered lanterns, battery chargers, radios, hot water heaters, laptop chargers, bicycle-powered generators, etc.

9. Consider converting your vehicle to biodiesel. This will provide you with more flexibility as gas prices become prohibitively expensive. As explained in "Part III. Alternatives to Oil: Fuels of the Future or Cruel Hoaxes?," biodiesel is not a scalable alternative for the entire civilization. It does, however, have value as a boutique fuel for those of us in the Peak Oil community. One of the advantages of biodiesel over gasoline is that you can also stockpile large amounts of biodiesel with relative ease for the inevitable shortages. In contrast it's very difficult to safely stockpile large quantities of gasoline. (It may not even be legal; I don't know.)

10. Consider taking an organic farming class or joining a local food cooperative. We all need to learn about soil and non-oil-powered agriculture. The Website communitysolutions.org is an excellent place to start.

11. Begin learning basic emergency medical procedures.

12. Investigate alternative forms of health care such as bioenergetic healing, self-hypnosis, etc. Pharmaceutical-based health care will soon become too expensive for anybody but the super-wealthy.

13. Reduce your debt load as much as possible.

14. Begin thinking how you are going to survive through blackouts, food/water shortages, and economic breakdowns.

15. Begin educating yourself about monetary reform and local currencies. Two good places to start are localcurrency.org and solari.com.

16. Begin accepting death. Let's not sugarcoat the situation: between the combined effects of international financial insolvency, global climate change, and energy famine, you can pretty much kiss your ass goodbye.

    Even if you survive, you will witness an unprecedented amount of death and suffering during the later stages of the oil crash. You may as well begin preparing now.

17. Get into some type of regular exercise program. When I found out about Peak Oil, I initially got so depressed I stopped lifting weights, which had been a lifelong passion of mine. After losing 30 pounds and feeling like total crap I finally got back into my program. I've

found that it really helps me deal with the situation in an upbeat fashion.

18. Develop a sense of humor. As you can tell, I have not let the fact I spend my days researching and writing about the "end of the world" dampen my sense of humor. The future is not going to be pleasant. A sense of humor will make it much easier to deal with.

108. Would it be a good time to look into buying a solar-powered home, if I have the financial resources to do so?

George W. Bush sure seems to think so. He has a state-of-the-art solar-powered, "off-the-grid" home described as an "environmentalist's dream home." That should tell you something.

109. How am I supposed to help stop the military-industrial complex that seems to have taken over the world?

Are you ready to be a truly revolutionary American and put down your wallet? The military-industrial complex has taken over because we've given it our money, mostly for useless items that we don't need. Limit your consumer purchasing as much as you can and you will do more to slow down, and perhaps stop, the military-industrial complex than you will ever do by attending a peace march. Marching for peace does nothing to address the true cause of our problems. Driving to the march and stopping at Starbucks on the way is actually making the situation worse. All marching for peace does is waste precious time and resources, while giving the "powers that be" the opportunity to deny we are creeping towards fascism.

I used to participate in peace marches but once I learned about Peak Oil I stopped. I felt like a bit of a hypocrite protesting an oil war when my lifestyle was every bit as energy-intensive and oil-powered as the lifestyles led by most of the pro-war folks. Ironically, the folks at the protests who use the least amount of oil per day are most often the police, many of who are dispatched on bicycles.

It's simple: the best way to stop the military-industrial complex is to pull your money out of the market and stop buying useless crap, especially new cars!

Additionally, do not let anybody you care about enlist in the military or allow themselves to be drafted. Check the Website www.objector.org for more information. If you are of draft age, begin working on your conscientious objector status right now. The loopholes of the Vietnam era that allowed people to escape to Canada or get student deferments have been closed.

110. How am I supposed to maintain a positive mental attitude now that I know industrial civilization is about to collapse? How should I prepare emotionally?

First of all, you're going to have to give yourself some time to deal with this. You just found out civilization is about to collapse. Unless you're already thoroughly medicated, you're going to be shaken up a good deal. For most people, there seems to be a certain "Post-Peak-Oil Depression" period, ranging in time from 6 weeks to 6 months. There's really no way to skip or go around this adjustment period. This is perfectly normal and to be expected.

In effect, you are coming to grips with the very essence of mortality. This is not surprising when you consider the amazing similarities between the life of an oilfield and the life of a human being:

1. When an oilfield is discovered, it is typically a time of great celebration and happiness for those who found it. The same is true for most parents when they have a baby.

2. After the initial discovery of an oilfield, much work must be done by those who discovered it. They have to invest all sorts of time and money for the oilfield to reach its potential. The same is true for the parents of a young child.

3. For many years, the oilfield will become increasingly productive. The same is true for a human being. From the time they are born to about middle age, each year (on average) will be more productive than the last.

4. Once an oilfield is 50 percent depleted, it reaches its "peak" of productivity. Likewise, most human beings reach the peak of their productivity around middle age.

5. Once an oilfield passes its peak, it begins a decline. Eventually, it requires more energy than it produces. At that point, the field is abandoned as it is essentially dead. The same is true for a human being.

There are also some interesting similarities between civilizations and human beings:

1. Every great civilization has collapsed. Every human being has died.

2. Every civilization has had good things and bad things about it. Likewise, every human being has good characteristics and bad characteristics.

Personally, I've found it's important to keep a bit of perspective about this whole situation. To illustrate, I want you to say the following statement out loud, three consecutive times:

"Because of Peak Oil, I am incredibly
depressed. Life is no longer worth living."

Bet that made you feel like warmed-over crapola, didn't it? Now, say the following statement out loud, three consecutive times:

"Because of Peak Oil, I am motivated to make
the most of the time and resources available to me."

Doesn't that feel way better than the first statement? Since both statements are equally true, you may as well believe and focus on the one that gets you into a more productive, upbeat state of mind.

The fact is you live in a society where you could die a horrible death on any given day. In the US, 40,000 people are killed and 3 million are injured every year in auto accidents. It's quite likely that future generations will regard automobiles in much the way we regard dinosaurs: as giant, scary, deadly things that used to roam around maiming and killing other living creatures. It's rather ironic when you consider where oil comes from.

I think it's also important to keep in mind that while our way of life may be the only way of life we have ever known, it is certainly not an optimal way of life. We live in a civilization where, for instance, a "successful" person is somebody who acquires tons of debt to earn a degree which will allow them to obtain a high-paying but extremely stressful job so they can purchase a big car in order to drive two hours a day to the job so they can pay for the gas to drive to the job so they pay off the debt which they acquired in order to earn the degree to obtain the job so they could purchase the car to drive to the job so they could pay for the gas which their neighbor's 19-year-old son just got his head blown off in Iraq for.

That's a pretty fucked-up way of organizing a civilization, don't you think? It's hard to imagine we'll replace it with anything worse.

111. I tried to tell my relatives about this, but they refused to listen. How can I get them to take this issue seriously?

Other than giving them this book, the film *End of Suburbia*, or perhaps one of the other fine books recently published about Peak Oil, there isn't much you can do. People who don't want to hear this information just aren't going to listen. It won't matter who the messenger is, what their credentials are, or how many facts are brought to the table.

My best advice to you is to stop wasting your time on a lost cause. There will be people living in abandoned gas stations 25-50 years from now, still naïvely insisting on things like, "Peak Oil is a myth propagated by the oil companies to drive up the price of oil," or, "As soon as we drill ANWR, the problem will be solved," or, "If the oil companies hadn't suppressed alternative energy technologies, this would never have happened," and other such delusional nonsense.

Unfortunately, the fact that many (if not most) people won't accept this information until it's too late is something we just have to accept and deal with as best we can.

If it's any consolation, I can't get many of my relatives to listen either! In fact, one relative threw a copy of the first edition of this book at another relative in the middle of a crowded restaurant and proclaimed, "I don't want to hear any doom and gloom!"

112. Why do you think this is happening? What do you think the "big picture" is?

Some evolutionary biologists believe that whenever an ecosystem has an excess of a particular resource, a species will arise to make use of that resource. Over time, this new species will exhaust the resource they were adapted to make use of. This brings the system back into balance. At that point, the species has fulfilled its evolutionary purpose. It will then go extinct. According to biologist David Price:

> Before the appearance of Homo sapiens, energy was being sequestered more rapidly than it was being dissipated. Then human beings evolved, with the capacity to dissipate much of the energy that had been sequestered, partially redressing the planet's energy balance. The evolution of a species like Homo sapiens may be an integral part of the life process, anywhere in the universe it happens to occur. If organic energy is sequestered in substantial reserves, as

geological processes are bound to do, then the appearance of a species that can release it is all but assured. Such a species, evolved in the service of entropy, quickly returns its planet to a lower energy level. In an evolutionary instant, it explodes and is gone.[280]

My take on Mr. Price's theory is this: Many eons ago, God was spending a leisurely morning in Her office when one of the angels, perhaps Michael or Gabriel, walked in. The angel said to God:

God, we got a problem down on Earth. You see, all the energy from the Sun has been accumulating inside the Earth as this black gooey stuff we decided to call oil. This wasn't a problem for the first few billion years, but now it looks like Earth has more stored energy than it can handle. Me and the other angels tried to fix the problem, but so far we haven't been able to figure anything out.

God sat in Her chair and thought for a moment. Then, in what amounted to the ultimate "Eureka" moment, God jumped out of Her seat and exclaimed, "I know what I'll do! I'll create a species that's dumb enough to use the stuff!"

A few billion years later, we're pretty close to accomplishing our original purpose. As a result, we stand on the brink of extinction. Hopefully, if we show God we are capable of doing something other than just consuming oil, she'll find a new purpose for us.

113. As things begin to collapse, do you think society will finally make good on Shakespeare's admonition to "kill all the lawyers"?

Oh shit.

About the Author

I was born and raised in California. I received my undergraduate degree in Political Science from the University of California at Davis in June 2000. I received my law degree from the University of California at Hastings College of the law in May 2003. In December 2003, I received my license to practice law in the state of California.

I learned about Peak Oil shortly after taking the Bar exam and have dedicated myself to dealing with.

If you would like to contact me, send an email to matt@lifeaftertheoilcrash.net

Endnotes and Sources

1   See the Association for the Study of Peak Oil and Gas, online at http://www.peakoil.net

2   Dr. David Goodstein, author of *Out of Gas: The End of the Age of Oil*, is a professor of physics and the vice-provost of Caltech University. According to Goodstein, Peak Oil is a grave concern. See Caltech Press Release (January 20, 2004). Archived at http://pr.caltech.edu/media/Press_Releases/PR12478.html

3   Matt Simmons, founder of the investment bank Simmons and Company International, is one of the most successful and respected energy investment bankers in the world. He has written and spoken at length about the severe ramifications of Peak Oil. For an online archive of his papers and speeches see: http://www.simmonsco-intl.com/research.aspx?Type=msspeeches

4   Michael Ruppert, "Revealing Statements from a Bush Insider about Peak Oil and Natural Gas Depletion," *From the Wilderness Publications,* (June 12, 2003). Archived at: http://www.fromthewilderness.com/free/ww3/061203_simmons.html

5   Matthew Simmons, "Energy in the New Economy: Limits to Growth," Presented at the Energy Institute for the Americas; Oklahoma City, Oklahoma (October 2, 2000). Archived at http://www.simmonscointl.com/files/79.pdf

6   White House Press Release (May 3, 2001). Archived at http://www.whitehouse.gov/news/releases/2001/05/20010503-4.html

7   Michael C. Ruppert, "Colin Campbell on Oil," *From the Wilderness Publications,* (October 23, 2002). Archived at http://www.fromthewilderness.com/free/ww3/102302_campbell.html

8   To study the oil production bell curves for 42 oil-producing nations, see the graphs based upon data presented by Richard Duncan and Walter Youngquist, in a paper entitled "The World Petroleum Life Cycle." Archived at http://dieoff.com/page133.htm

9   James Mackenzie, "Plausible Scenarios for Future Global Oil Production," Excerpted from *Oil as a Finite Resource: When is Global Oil Production Likely to Peak?,* Archived at http://pubs.wri.org/pubs_content_text.cfm?ContentID=382

[10] See estimates given by members of the Association for the Study of Peak Oil and Gas: http://www.peakoil.net. See also the index to the experts on Peak Oil available at http://www.hubbertpeak/experts/

[11] See Richard Heinberg, "The Petroleum Plateau," *Museletter* (Number 135/May 2003). Archived at http://www.museletter.com/archive/135.html

[12] "BP Acquiring Arco for 25.6 Billion," *Alexander's Gas and Oil Connections,* (January 4, 1999). Archived at http://www.gasandoil.com/goc/features/fex91841.htm

[13] Jon Thompson, "A Revolutionary Transformation," *The Lamp,* Volume 85, No. 1. Archived at: http://www2.exxonmobil.com/corporate/newsroom/publications/thelampnol_2003/page_5.html

[14] The Saudi saying was brought to my attention by Richard Heinberg in his book, *The Party's Over: Oil, War, and the Fate of Industrial Civilizations* on p. 81

[15] Ted Trainer, "The Death of the Oil Economy," *Earth Island Journal,* (Spring 1997). Archived at http://dieoff.com/page116.htm

[16] Dale Allen Pfeiffer, "Eating Fossil Fuels," *From the Wilderness Publications* (October 3, 2003), citing David Pimentel and Mario Giampetro, "Food, Land, Population and the US Economy," *Carrying Capacity Network,* (November 21, 1994). Archived at http://www.fromthewilderness.com/free/ww3/100303_eating_oil.html

[17] According to a CIA study cited by John Fawcett Long, "The Farm Bill Gets Down on the Farm," *Washington Free Press* (April/May 1995). Archived at http://www.washingtonfreepress.org/15/Farm.html

[18] Dr. Allan R. Hoffman, "The Connection: Water and Energy Security," *Energy Security* (August 13, 2004). Archived at http://www.iags.org/n0813043.htm

[19] Caryl Johnston "Modern Medicine and Fossil Fuel Resources," Archived at http://mysite.verizon.net/vze495hz/id19.html

[20] David R. Klein, "The Introduction, Increase, and Crash of Reindeer on St. Matthew Island," (1966). Archived at http://dieoff.org/page80.htm

21 As quoted by Jared Diamond, "Easter Island's End," *Discover Magazine,* (August 1995) Archived at: http://www.hartford-hwp.com/archives/24/042.html

22 Ibid.

23 Ibid.

24 Ibid.

25 Clive Ponting, "Lessons of Easter Island," excerpted from *A Green History of the World* (Penguin Books, 1992) p.168-170; Archived at http://www.eco-action.org/dt/eisland.html

26 Ibid.

27 Ibid.

28 Jared Diamond, "Twilight at Easter," *The New York Review of Books,* (March 25, 2004). Archived at http://www.nybooks.com/articles/16992

29 See note 21

30 Ibid.

31 David Price, "Energy and Human Evolution," *Population and Environment, A Journal of Interdisciplinary Studies,* Volume 16, No. 4 p. 301-319, (March 1995). Archived at http://www.dieoff.com/page 137htm#2

32 David Pimentel and Mario Giampetro, "Food, Land, Population, and the US Economy," *Carrying Capacity Network,* (November 21, 1994).

33 As quoted by William Engdahl, "Iraq and the Problem of Peak Oil," *Current Concerns,* (January 26, 2004). Archived at http://www.currentconcerns.ch/archive/2004/01/20040118.php; To read the full text of Dick Cheney's speech, see http://www.peakoil.net/Publications/Cheney_PeakOil_FCD.pdf

34 As quoted by Larry Everest, "Saddam, Oil and Empire: Supply and Demand," *Counterpunch* (December 13-14, 2003). Archived at http://www.counterpunch.org/everest12132003.html

35 As quoted by Ronald Bailey, "Energy Independence: An Ever Receding Mirage," *Reason Online* (July 21, 2004). Archived at http://www.reason.com/rb/rb072104.shtml

36 Michael C. Ruppert, "Behind the Blackout: Bush Insider Speaks," *From the Wilderness*, (August 21, 2003). Archived at http://www.fromthewilderness.com/free/ww3/082103_blackout_summary.html

37 Adam Porter, "Is the World's Oil Running Out Fast," *British Broadcasting Company* (June 7, 2004). Archived at http://news.bbc.co.uk/1/hi/business/3777413

38 James Robbins, "Is Saudi Oil Worth the Trouble?" *National Review Online*, (June 2, 2004). Archived at http://www.nationalreview.com/robbins/robbins200406020835.asp

39 Dr. Richard C. Duncan, "The Peak of World Oil Production and the Road to the Olduvai Gorge," (November 13, 2000). Archived at http://www.dieoff.com/page224.htm

40 Chris Skrebowski "Over a Million Barrels of Oil Lost Per Day Due to Depletion," *Petroleum Review*, (August 24, 2004).

41 Ibid.

42 Graham James, "World Oil and Gas Running Out," *CNN International*, (October 2, 2003). Archived at http://edition.cnn.com/2003/WORLD/europe/10/02/global.warming/

43 Joseph A. Giannone, "Major Oil Stocks Fall as Shell Revises Reserves," *Reuters Business News*, (January 9, 2004). Archived at http://uk.biz.yahoo.com/040109/80/eiqr3.html

44 "El Paso Trims Reserves 41 Percent, *Reuters* (February 13, 2004).

45 "Shell Trims Its Reserves Again," *British Broadcasting Company*, (May 25, 2004). Archived at http://news.bbc.co.uk/1/hi/business/374287.stm

46 "Shell Admits Fueling Corruption," *British Broadcasting Company*, (June 11, 2004). Archived at http://news.bbc.co.uk/1/hi/business/3839679.stm

47 "Shell CEO Gets Buyout," *British Broadcasting Company*, (June 25, 2004). Archived at http://news.bbc.co.uk/1/hi/business/3839679.stm

48 Tim Webb, "BP May be Forced to Downgrade Reserves," *The Independent* (June 4, 2004)

[49] As quoted in "What If?," *The Economist* (May 27, 2004).

[50] Ellen Read, "Financial Fallout Hardly Noticed," *New Zealand Herald* (August 4, 2004).

[51] "This Week in Petroleum," Archived at http://tonto.eia.gov/oog/info/twip/twip.asp

[52] "Tear Gas, Protests Greet Indonesia's Fuel Price Increase," *CNN International* (June 16, 2001).

[53] Jimmy Langman, "Bolivia's Protests of Hope," *The Nation,* (October 22, 2003). Archived at http://www.thenation.com/doc.mhtml?i=20031103&s=langman

[54] "California Truckers Protest Fuel Prices," *BizJournal* (April 28, 2004). Archived at http://www.bizjournals.com/losangeles/stories/2004/04/26/daily37.html

[55] Carolyn Said, "Burning up Over Gas Prices," *San Francisco Chronicle,* (June 9, 2004).

[56] Mark Townsend and Martin Bright, "Army Guard on Food if Fuel if Fuel Crisis Flares," *The Guardian,* (June 6, 2004). Archived at: http://www.guardian.co.uk/oil/story/0,11319,1232445,00.html

[57] Dean E. Murphy, "Farm Fuel Thefts Rise Along with Prices," *The New York Times,* (July 12, 2004).

[58] "China Farmers Get Death for Oil Theft," *The San Jose Mercury News,* (July 25, 2004).

[59] Jay Apt and Lester Lave, "US Blackouts Inevitable," *The Washington Post* (August 10, 2004). Archived at http://www.energybulletin.net/1591.html

[60] Jonathan Watts, "China's Growth Flickers to a Halt," *The Guardian,* (July 4, 2004). Archived at http://news.bbc.co.uk/2/hi/asia-pacific/3916789.stm

[61] "Beijing Brown Out to Save Power," *BBC News,* (July 22, 2004). Archived at http://news.bbc.co.uk/2/hi/asia-pacific/39116789.stm

[62] Ibid.

63   David Good, "Will Shortfalls in World Grain Production Continue?" *Farmdoc*, (November 17, 2003). Archived at http://www.farmdoc.uiuc.edu/marketing/weekly/html/111703.html

64   "China's Rising Grain Prices Could Signal Global Food Crisis," Archived at http://www.terradaily.com/2003/031119092535.t70a5roc.html

65   Richard Heinberg, "Oil and Gas Update," *Museletter* Number 142 (January 2004)

66   Wayne Wenzel, "Farm Bureau Sounds Alarm on Natural Gas," *Farm Industry News* (September 22, 2004). Archived at http://farmindustrynews.com/news/Farm-Bureau-natural-gas-092204/

67   Chris Skrebowski "Over a Million Barrels of Oil Lost Per Day Due to Depletion," *Petroleum Review,* (August 24, 2004).

68   Lita Epstein, C.D. Jaco, and Julianne C. Iwersen-Neimann *The Politics of Oil* p. 255, citing the National Resource Defense Counsel's report, "Energy Policy for the Twenty-First Century." Penguin Group (2003)

69   "ANWR Oil Would Have Little Impact," *MSNBC,* (March 16, 2004) Archived at http://www.msnbc.msn.com/id/4542853/

70   Dale Allen Pfeifer, "Much Ado About Nothing: Whither the Caspian Riches," *From the Wilderness Publications* (December 5, 2002). Archived at http://www.fromthewilderness.com/free/ww3/120502_caspian.html

71   Richard Heinberg, *The Party's Over: Oil, War and the Fate of Industrial Civilizations* p. 112 citing Dr. Walter Youngquist, *Geodestinies,* p. 216

72   "Petro-Canada Reviews Oil Sands Strategy," *Rigzone*, (May 2, 2003). Archived at http://www.fromthewilderness.com/free/ww3/062303_nat_gas_crisis.html

73   As quoted by Tim Appenzeller, "The End of Cheap Oil," *National Geographic*, (June 2004) p. 105

74   Ibid.

[75] John Attarian, "The Coming End of Cheap Oil," *The Social Contract* (Summer 2002), citing both: "New Sources as Oil Supplies Grow, U.S. Is Less Reliant on the Middle East," *Wall Street Journal*, (March 15, 2002) and Dr. Colin Campbell, *ASPO-ODAC Newsletter* No. 15, p. 11 (March 2002).

[76] Richard Heinberg, *The Party's Over: Oil, War and the Fate of Industrial Civilizations* p. 115 citing Energy Information Agency, *Annual Energy Outlook* 1998 with Projections to 2020, p. 17

[77] Dr. Colin Campbell, "Misleading USGS Report," *Solar Quest Net News Service* (March 25, 2000). Archived at http://www.hubbertpeak.com/news/article.asp?id=848

[78] See Note 76 Heinberg at page 116 citing Howard T. Odum and Elisabeth C. Odum, *A Prosperous Way Down*, p. 169.

[79] Matthew R. Simmons, "The World's Giant Oilfields," M. King Hubbert Center for Petroleum Supply Studies, Colorado School of Mines, (January 2002). Archived at http://hubbert.mines.edu/news/Simmons_02-1.pdf

[80] Ibid

[81] C.J Campbell and Jean Laherrere, "The End of Cheap Oil?" *Scientific American*, (March 1998). Archived at http://www.dieoff.org/page40.htm

[82] Chris Skrebowski, "Oil Field Mega Projects, 2004." *Petroleum Review* (January 2004).

[83] Carola Hoyos, "A Battle Against Shrinking Reserves," *Financial Times,* (January 7th, 2004)

[84] John Attarian, "The Coming End of Cheap Oil," *The Social Contract*, (Summer 2002), citing Albert A. Bartlett, "An Analysis of US and World Oil Production Patterns using Hubbert-Style Curves," *Mathematical Geology*, vol. 32, no. 1 (January 2000)

[85] Michael Ruppert, "Colin Campbell on Oil," *From the Wilderness,* (October 10, 2002) Archived at http://www.fromthewilderness.com/free/ww3/102302_campbell.html

[86] Dr. Colin Campbell, *Association for the Study of Peak Oil Newsletter,* no. 35 (November 2003). Archived at http://www.asponews.org/HTML/Newsletter35.html

[87]   Tim Macalister, "BP Should Consider Mother of All Mergers," *The Guardian* (July 15, 2004). Archived at:
http://www.guardian.co.uk/business/story/0,3604,1261452,00.html

[88]   As quoted by Dr. Colin Campbell, "The Imminent Peak of Global Oil Production," (March 2000). Archived at
http://www.feasta.org/documents/feastareview/Campbell.htm

[89]   "Industry Comments," *The Energists* (June 1, 2000). Archived at
http://www.energists.com/industry_comments.html

[90]   Ibid.

[91]   "Wrong Sizing," *Oil and Gas Investor* (January 2001). Archived at
http://www.oilandgasinvestor.com/previous/0101/completions.htm

[92]   "Oil Industry Waits for Skilled Workers," *Alexander's Gas and Oil Connections,* (January 7, 2001). Archived at
http://www.gasandoil.com/goc/features/fex/13163.htm

[93]   Ibid.

[94]   Ibid.

[95]   Richard Heinberg, *Museletter* (Number 142/January 2004) citing Richard Duncan, "Three World Oil Forecasts Predict Peak Oil Production," *Oil and Gas Journal*, May 26, 2003 p. 18-21

[96]   Chris Skrebowski "Over a Million Barrels of Oil Lost Per Day Due to Depletion," *Petroleum Review,* (August 24, 2004).

[97]   Jay Schempf, "Matt Simmons Hopes He's Wrong," *Petroleum News* (August 1, 2004). Archived at
http://www.petroleumnews.com/pnads/238338932.shtml

[98]   Hamish Mcrae, "Crisis, Crunch, or Crumble, It's Time We Started Fearing for the Future of Saudi Oil," *The Independent* (August 1, 2004). Archived at
http://news.independent.co.uk/business/comment/story.jsp?story=546690

[99]   Dr. Ali Samsam Bakhtiari, "Middle East Oil Production to Peak Within the Next Decade," *Oil and Gas Journal,* (July 7, 2003) p. 24, as cited by G.R. Morton, "Trouble in the World's Largest Oil Fields," Archived at
http://home.entouch.net/dmd/ghawar.htm; See also, Bakhtiari's statements as reported by Michael C. Ruppert, "Paris Peak Oil

Conference Reveals Deepening Crisis," *From the Wilderness Publications* (June 9, 2003). Archived at http://www.fromthewilderness.com/free/ww3/053103_aspo.html

[100] Julian Darley, "Is Saudi Arabia Still the King of Oil?," *Alternet* (April 29, 2004). Archived at http://www.alternet.org/story/18555

[101] Adam Porter, "Is the World's Oil Running out Fast?" *BBC News Online* (June 7, 2004). Archived at http://news.bbc.co.uk/1/hi/business/3777413.stm

[102] Julian Darley, "A Tale of Two Planets," *From the Wilderness Publications,* (March 17, 2004). Archived at http://www.fromthewilderness.com/free/ww3/031704_two_planets.html

[103] Ibid

[104] Massod Farivar, "NYMEX Crude Resumes Rally, Yukos Concerns," *E-Commerce Times,* (August 17, 2004). Archived at http://groups.yahoo.com/group/energyresources/message/61754

[105] Jonathan Prynn and Sarah Marks, "Oil Price Threat to All of Economy," *This is London,* (June 1, 2004). Archived at http://www.thisislondon.co.uk

Patrice Hill, "Oil Prices Surge After Terror Attack in Khobar," *The Washington Times,* (June 2, 2004). Archived at http://www.washtimes.com/business/20040602-010024-8060r.htm

[107] Mark Townsend and Paul Harris, "Now the Pentagon Tells Bush: Climate Change Will Destroy Us," *The Guardian* (February 22, 2004). Archived at http://observer.guardian.co.uk

[108] Ibid.

[109] David Adam, "Oil Chief: My Fears for the Planet," *The Guardian,* (June 17, 2004). Archived at http://www.guardian.co.uk/climatechange/story/0,12374,1240566,00.html

[110] Dr. Kenneth Watt, "Book Review of *Geodestinies: The Inevitable Control of Earth Resources Over Nations and Individuals* by Dr. Walter Youngquist." Archived at http://www.energycrisis.com/youngquist/

[111] Michael Kane, "Beyond Peak Oil," *From the Wilderness Publications,* (September 17, 2004)

[112] Ibid. citing Stuart Baird, "The Automobile and the Environment," Archived at http://www.iclei.org/EFACTS/AUTO.HTM

[113] "Computer Manufacturing Soaks Up Fossil Fuels," *UN News Service,* (March 8, 2004). Archived at http://www.undp.org.vn/mlist/ksdvn/032004/post9.htm

[114] Richard Heinberg, *The Party's Over: Oil, War and the Fate of Industrial Civilizations* (New Society Publishers 2003) p.167

[115] Bruce Thomson, "The Oil Crash and You," Archived at http://www.culturechange.org/issue18/oilcrash.html

[116] Ibid.

[117] Dr. David Goodstein, *Out of Gas: The End of the Oil Age,* (2004). As excerpted by Lee Dye, "Dwindling Oil Supplies to Bring Crisis," *ABC News* (February 11, 2004). Archived at http://abcnews.go.com/sections/SciTech/DyeHard/oil_energy_dyehard_0 40211-1.html

[118] Dr. Paul B. Weisz, "Basic Choices and Constraints on Long Term Energy Supplies," *Physics Today,* (July 2004). Archived at http://www.physicstoday.org/vol-57/iss-7/p47.html#tab1

[119] John Gever et al., *Beyond Oil: The Threat To Food and Fuel in the Coming Decades* (1991) p.67, as cited by Jay Hanson, "Fossil Gate," Archived at http://www.dieoff.com/page122.htm

[120] Energy Information Agency

[121] Bruce Thomson, "The Oil Crash and You," Archived at http://www.culturechange.org/issue18/oilcrash.html

[122] Ibid.

[123] Ibid.

[124] Paul Rincon, "Ocean Methane Stocks Overstated," *BBC News Online,* (February 17, 2004). Archived at http://news.bbc.co.uk/1/hi/sci/tech/3493349.stm

[125] Ibid.

[126] Richard Heinberg, *The Party's Over: Oil, War and the Fate of Industrial Civilizations* (New Society Publishers 2003) p.151.

[127] Ibid.

[128] It doesn't matter how advanced technology becomes: it will never carry more energy than it took to obtain it due to the Second Law of Thermodynamics which dictates that a process will "output" less energy than was "input" into the process.

[129] Bruce Thomson, "The Oil Crash and You," Archived at http://www.culturechange.org/issue18/oilcrash.html

[130] Bill Moore, "Two Hundred Hours," *EV World*, (February 22, 2004). Archived at: http://groups.yahoo.com/group/energyresources/message/52194

[131] Dom Crea, "Drunk on Hydrogen," *The Rant*, (February 24, 2003). Archived at http://www.therant.info/archive/000519.html

[132] Joseph Romm, "The Hype About Hydrogen: Fact and Fiction in the Race to Save the Climate," Archived at http://www.culturechange.org/hydrogen.htm

[133] Donald Barlett and James Steele, "Why America is Running Out of Gas," *Alexander's Gas and Oil Connections*, (August 8, 2003).

[134] Ibid.

[135] Energy Information Agency

[136] Richard Heinberg, *The Party's Over: Oil, War and the Fate of Industrial Civilizations* (New Society Publishers, 2003) p. 138-139

[137] Dr. David Goodstein, author of *Out of Gas The End of the Oil Age* puts the number at 25 years. Richard Heinberg, author of *The Party's Over* puts the number at 40 years. Either way, combined with its other shortcomings and obstacles, nuclear energy isn't going to save us.

[138] In 2002, the US consumed over 97 Quadrillion BTUs of energy. Only .064 of these came from solar power. .064 divided by 97 = .000657. See http://www.eia.doe.gove/aer/txt/ptb0101.html and http://www.eia.doe.gov/emeu/aer/txt/ptb0103.html

139 Paul Roberts, *The End of Oil: On the Edge of a Perilous New World* Houghton Mifflin Company (2004) p. 202

140 As quoted by Brain Braiker, "Crude Awakening," *MSNBC* (February 17, 2004). Archived at http://msnbc.msn.com/id/4287300/

141 Theodore Butler, *Silver Profits in the New Century*, Available from http://www.investmentrarities.com

142 As quoted by John Gartner, "Up and Atom," *Alternet* (September 8, 2004). Archived at http://www.alternet.org/envirohealth/19812/

143 Bruce Thomson, "The Oil Crash and You," Archived at http://www.culturechange.org/issue18/oilcrash.html

144 In 2002, the US consumed about 97 BTUs of energy. Only .106 came from wind. .106 divided by 97 = .001. See: http://www.eia.doe.gov/emeu/aer/txt/ptb0101.html and http://www.eia.doe.gov/emeu/aer/txt/ptb0103.html

145 As quoted by Lee Dye, "Old Policies Make Shift From Foreign Oil Tough," *ABC News*, (July 28, 2004).

146 Lee Raymond, "Facing Some Hard Truths About Energy," *Petroleum World*, (June 12, 2004). Archived at http://www.energybulletin.net/newswire.php?id=624

147 For an example of what I'm talking about, see: Stewart Truelson, "Former CIA Chief Backs Renewable Fuels," *Farm Bureau Focus* (November 11, 2002). Archived at http://www.fb.org/views/focus/fo2002/fo1111.html

148 Dr. Walter Youngquist, "The Post-Petroleum Paradigm," *Population and the Environment: A Journal of Interdisciplinary Studies,* Volume 20, Number 4, (March 1999). Archived at http://www.dieoff.com/page171.htm

149 See http://en.wikipedia.org/wiki/Thermal_depolymerization

150 Jay Hanson, "Five Fundamental Errors," Archived at http://www.dieoff.com/page241.htm

151 Ibid.

¹⁵² As quoted by Dr. Albert Barlett, "The New Flat Earth Society," Archived at http://www.hubbertpeak.com/bartlett/flatEarth.htm

¹⁵³ Stephen Hawking, *The Universe in a Nutshell,* p. 158

¹⁵⁴ John Gever, *Beyond Oil: The Threat to Food and Fuel in the Coming Decades,* p. 67

¹⁵⁵ Colin Campbell, "Association for the Study of Peak Oil," No. 27 (March 2003). Archived at http://www.asponews.org/ASPO.newsletter.027.php

¹⁵⁶ Dale Allen Pfeiffer, "Drawing Lessons From Experience: The Agricultural Crisis in North Korea," *From the Wilderness* (November 17, 2003). Archived at http://www.fromthewilderness.com

¹⁵⁷ "Scores of North Korean Children Dead," *CNN News,* (April 8, 1997). Archived at http://www.cnn.com/WORLD/9704/08/korea.food/index.html

¹⁵⁸ "Kerry Aims to Reduce Foreign Oil Reliance," *Associated Press* (August 6,2004). Archived at http://abcnews.go.com/wire/Politics/ap20040806_706.html

¹⁵⁹ As quoted by Derrick Jackson, "Chasing the SUV Vote," *The Boston Globe,* (August 18, 2004). Archived at http://www.commondreams.org/views04/0818-02.htm

¹⁶⁰ White House Press Release (May 3, 2001). Archived at http://www.whitehouse.gov/news/releases/2001/05/20010503-4.html

¹⁶¹ See http://www.cbo.gov/ftpdocs/52xx/doc5241/doc40.pdf

¹⁶² See Note 164.

¹⁶³ David Flickling, "World Bank Condemns Defense Spending," *The Guardian,* (February 14, 2004). Archived at http://www.guardian.co.uk/globalisation/story/0,7369,1147888,00.html

¹⁶⁴ As quoted by David Sirota, Christy Harvey and Judd Legum "Red Planet Profits," *Center for American Progress: Daily Progress,* (January 15, 2004). Archived at http://www.tompaine.com/feature2.cfm/ID/9774

¹⁶⁵ Dr. Helen Caldicott, *The New Nuclear Danger*, p. 114, quoting William B. Scott, "USSC Prepares for Future Combat Missions in Space," *Aviation Week and Space Technology*, (August 5, 1996)

166   As quoted by David Sirota, Christy Harvey and Judd Legum, "Red Planet Profits," *Center for American Progress: Daily Progress,* (January 15, 2004). Archived at http://www.tompaine.com/feature2.cfm/ID/9774

167   Theresa Hitchens, "Reining in Our Weaponry: Is US Air Force Lost in Space?" *The San Francisco Chronicle,* (March 15, 2004). Archived at http://www.commondreams.org/views04/0315-05.htm

168   Julie Wakefield, "Moon's Helium-Three Could Power Earth," *Space.com* (June 30, 2000). Archived at http://www.space.com/scienceastronomy/helium3_000630.html

169   Ibid.

170   Satyabrata Rai Chowdhuri, "An Energy Source That's Out of This World," *The Asia Times*, (November 15, 2003). Archived at http://www.atimes.com/atimes/Global_Economy/EK15Dj01.html

171   As quoted by Michael Klare, "Bush-Cheney Energy Strategy: Procuring Other People's Oil," *Foreign Policy in Focus* (January 2004), Archived at http://www.fpif.org/papers/03petropol/politics.html

172   "Mexicans Outraged by Immigration-Oil Move," *Kansas City Star,* (May 10, 2003). Archived at http://www.kansascity.com/mld/kansascity/news/breaking_news/583334.htm

173   Peter Dale Scott, *Oil, War and Drugs*, p. 43.

174   Ibid, p. 65, citing http://www.narconews.com/pressbriefing21september.html

175   Michael Ruppert, "CIA, Drugs and Wall Street," *From the Wilderness,* (June 29, 1999) Archived at http://www.fromthewilderness.com/free/ciadrugs/dontblink.html

176   See Catherine Austin Fitts, "Narco Dollars for Dummies." Archived at: http://www.narconews.com/narcodollars1.html

177   Scott, *Oil, War and Drugs,* p.44 citing *Washington Post* (December 10, 2001).

178   Afghan Poppy Production Doubles," *CNN News*, (November 28, 2003). Archived at http://www.cnn.com/2003/ALLPOLITICS/11/28/afghanistan.drugs.reut/index.html

179 See generally, Peter Gorman, "Plan Columbia, The Pentagon's Shell Game," *From the Wilderness Publications* (March 10, 2003), Archived at http://www.fromthewilderness.com/free/ww3/033103_plan_columbia.html

180 As quoted by Catherine Austin Fitts, "Narco-Dollars for Dummies," Archived at http://www.narconews.com/narcodollars1.html

181 John Attarian, "The Coming End of Cheap Oil," *The Social Contract* (Summer 2002). Archived at http://thesocialcontract.com/cgi-bin/showarticle.pl?articleID=1094&terms=

182 Daniel Quinn, "Address to the Minnesota Social Investment Forum" (June 7, 1993). Archived at http://www.ishmael.com/Education/Writings/OnInvestments.shtml

183 David Ruppert, "US Military Wanted to Provoke War with Cuba," *ABC News,* (May 1, 2003) Archived at http://abcnews.go.com/sections/us/DailyNews/jointchiefs_010501.html

184. See "Pentagon Proposed Pretext for Cuba Invasion," *National Security Archive George Washington University.* Archived at http://www.gwu.edu/nsarchiv/news/20010430/

185 For a summary of Operation Northwoods, see at http://operation-northwoods.wikiverse.org/

186 George Monbiot, "America's Pipedream," *The Guardian* (October 23, 2001). Archived at http://www.guardian.co.uk/comment/story/0,3604,579071,00.html

187 Ibid.

188 Michael Klare, *Resource Wars: The New Landscape of Global Conflict* quoting a White House press statement dated August 1, 1997. Archived at http://www.thinkingpeace.com/Lib/lib062.html

189 As quoted by Lester W. Grou, "Hydrocarbons and a New Strategic Region: The Caspian Sea," *Military Review,* (May-June 2001). Archived at http://www.globalsecurity.org/military/library/report/2001/hydrocarbons.htm

190 "Afghanistan Plans Gas Pipeline," *BBC News Online,* (May 13, 2002). Archived at http://news.bbc.co.uk/1/hi/business/1984459.stm

191 Christian Berthelson, "New Scrutiny of Airlines Options Deals," *San Francisco Chronicle,* (September 19, 2001).

[192] Michael C. Ruppert, "Suppressed Details of Criminal Insider Trading Lead Directly to the CIA," *From the Wilderness,* (October 9, 2001). Archived at http://www.fromthewilderness.com/free/ww3/10_09_01_krongard.html

See also, "Mystery of Terror Inside Deals," *The Independent,* Archived at http://news.independent.co.uk/business/news/story.jsp?story=99402

[193] Ibid.

[194] Tom Flocco, "Profits of Death: Insider Trading and 9/11," *From the Wilderness* (December 6, 2001). Archived at http://www.fromthewilderness.com/free/ww3/12_06_01_death_profits_pt 1.html

[195] Christian Berthelson, "Suspicious Profits Go Uncollected/Airline Investors Seem to be Lying Low," *San Francisco Chronicle* (September 29, 2001). Archived at http://www.sfgate.com; See also, note 204

[196] FAA news release Aug/9/02; *Associated Press,* (August 13, 2002), quoted at http://www.globalresearch.ca/articles/THO311B.html

[197] "Payne Stewart Killed in Plane Crash," *Sports Illustrated,* (October 26, 1999).

[198] William Thomas, "Pentagon Says 9/11 Interceptors Flew Too Far, Too Slow, Too Late," *Global Research,* (November 19, 2003). Archived at http://www.globalresearch.ca/articles/THO311B.html

[199] Jon Dougherty, "Bill Allows Forced Vaccinations," *World Net Daily* (November 16, 2002). Archived at http://www.worldnetdaily.com/news/article.asp?ARTICLE_ID=29682

[200] Archived at http://www.thirdreich.net/Thought_They_Were_Free.html#Top

[201] Robert Freeman, "Will the End of Oil Mean the End of America?" *Commondreams.org,* (March 1, 2004). Archived at http://www.commondreams.org/views04/0301-12.htm

[202] Ibid.

[203] Nat Hentoff, "General Ashcroft's Detention Camp," *Village Voice,* (September 4-10, 2002). Archived at http://www.villagevoice.com/issues/0236/hentoff.php

[204] Wendy Balazik, "Former Senator Ashcroft: A Long Anti-Environmental Record," *Sierra Club News Release,* (January 9, 2001). Archived at http://www.commondreams.org/news2001/0109-08.htm

[205] For a thorough analysis of Henry Wallace's warnings as written in the *New York Times*, see: Thom Hartmann, "The Ghost of Vice-President Wallace Warns Us: 'It Can Happen Here.'" *CommonDreams.org* (July 19, 2004). Archived at http://www.commondreams.org/views04/0719-15.htm

[206] Clayton Cramer, "An American Coup d'Etat?" *History Today* (November 1995). Archived at http://www.claytoncramer.com/amcoup.html;

[207] Ibid.

[208] Ibid.

[209] http://www.carpenoctem.tv/cons/whitehouse.html

[210] See Cramer, Note 206

[211] Note 209 in addition to John L Spivak, "Wall Street's Fascist Conspiracy," *New Masses,* January 29, 1935, 9-15; February 5, 1935

[212] See http://members.tripod.com/~american_almanac/morgan4.htm

[213] As quoted by Mark Hand, "It's Time to Get Over It: John Kerry Tells the Antiwar Movement to Move On" *Press Action* (February 9, 2004). Archived at http://www.globalresearch.ca/articles/HAN403A.html

[214] As cited by Sonali Kolhatkar and James Ingalls, "Shattering Illusions: Kerry Doesn't' Need or Want Anti-War Activists," *ZNET,* (July 31, 2004). Archived at http://www.zmag.org

[215] As quoted by Bill Van Aucken, "Kerry, Edwards Vow to Continue War and Social Reaction," (July 31, 2004). Archived at http://www.wsws.org/articles/2004/jul2004/dnc-j31.html

[216] See Note 214

[217] Gan Golan, "Report From Boston: Stay Out of the Free Speech Zone," (July 26, 2004). Archived at http://nyc.indymedia.org/feature/display/98452/index.php

218 Michael Avery, "The Demonstration Zone at the Democratic National Convention," *Truthout* (July 25, 2004). Archived at http://www.truthout.org

219 Neil Mackay, "The West's Battle for Oil," *Sunday Herald,* (October 20, 2002). Archived at http://www.sundayherald.com/28224

220 "Defense Redefined Means Securing Cheap Energy," *Sydney Morning Herald,* (December 26, 2002). Archived at http://www.smh.com.au/articles/2002/12/25/1040511092926.html

221 Dr. Paul R. Ehrlich, *The Population Bomb*, 1st Edition (Ballantine Books, 1968). Prologue.

222 Ibid

223 "What is N.S.S.M 200 and Why Do Western Leaders Care So Much About Population Control?" Archived at http://www.africa2000.com/INDX/nssm200.htm; To see N.S.S.M 200 in its entirety, go to http://www.africa2000.com/SNDX/nssm200all.html

224 Ibid. quoting, "Global Demographic Trends to the Year 2010: Implications for US Security," *The Washington Quarterly* (Spring 1999).

225 Ibid. quoting, "Population Change and National Security," *Foreign Affairs* (Summer 1991); Excerpt of original paper archived at http://www.foreignaffairs.org/1991/3.html

226 http://www.globalpolicy.org/security/data/childdeaths.htm

227 Ibid.

228 Leah Wells, "Water Woes: In Iraq, Water and Oil do Mix," *CounterPunch.org* (May 16, 2003). Archived at http://www.counterpunch.org/wells05162003.html See also: Ghali Hassan, "Unmasked War Against Iraqi Children," *Counter Currents* (August 4, 2004). Archived at http://www.countercurrents.org/iraq-hassan040804.htm

229 Rajiv Chandrasekaran, "US Funds for Iraq Largely Unspent," *Washington Post,* (July 4, 2004). Archived at http://www.washingtonpost.com

230 Katherine Neuffer, "Iraqis Trace Surge in Cancer to US Bombings," *Boston Globe* (January 26, 2003). Archived at http://www.commondreams.org/headlines03/0126-03.htm

[231] Christpher Petherick, "Birth Defects Tied to GWS," *American Free Press*, (June 29, 2003). Archived at
http://www.americanfreepress.net/06_29_03/Birth_Defects_Tied/birth_defects_tied.html

[232] Scott Peterson, "Remains of Toxic Bullets Litter Iraq," *Christian Science Monitor*, (May 15, 2003). Archived at
http://www.commondreams.org/headlines03/0515-01.htm

[233] Bob Nichols, "There are no Words: Radiation in Iraq Equals: 250,000 Nagasaki Bombs" *Dissident Voice* (March 27, 2004).
Archived at http://www.dissidentvoice.org/Mar04/Nichols0327.htm

[234] As quoted by David Wildman, "Oceans of Greed," *The Weekly Dig*, Archived at http://www.weeklydig.com/dig/content/2372.aspx

[235] See generally, Paul Harris, "America's War Against Europe," *Yellow Times* (February 19, 2003). Archived at
http://www.yellowtimes.org/article.php?sid=1083

[236] See generally, William Engdahl, "A New American Century: Iraq and the Hidden Euro-Dollar Wars," (June 2003). Archived at
http://www.globalresearch.ca/articles/ENG401A.html

[237] See generally, notes 235 and 236

[238] "Iranian Leaders Threaten US" *World Net Daily*, (July 14, 2004) Archived at
http://www.worldnetdaily.com/news/article.asp?ARTICLE_ID=39430

[239] As quoted by Michael Ruppert, "The Bill Collector Calls," *From the Wilderness Publications* (June 21, 2004). Archived at
http://www.fromthewilderness.com/free/ww3/062104_berlin_peak.html

[240] Tanya Hsu, "Who Really Wants to Invade Saudi Arabia and Why," (July 4, 2004). Archived at http://www.irmep.org/essays/ksa.htm

[241] Ibid.

[242] As quoted by Robert Dreyfuss, "The Thirty Year Itch," *Mother Jones Magazine*, (March/April 2003), Archived at
http://www.motherjones.com/news/feature/2003/03/ma_273_01.html

[243] See Hsu, Note 240.

244   Karl B.Koth, "Why Should the United States Care Who is the President of Venezuela?" Archived at http://www.vheadline.com/readnews.asp?id=21779

245   Jonathan Katzenellenbogen, "United States Eyes West African Crude," *Business Day,* (December 10, 2002). Archived at http://www.worldpress.org/Africa/899.cfm

246   See "World Oil Transit Chokepoints," Energy Information Agency, Archived at http://www.eia.doe.gov/emeu/security/choke.html

247   See generally, Michael C. Ruppert, "Saudi Arabia, West Africa: Next Stops in the Infinite War for Oil," From the Wilderness Publications, (May 15, 2003). Archived at http://www.fromthewilderness.com/free/ww3/051503_saudi_africa.html

248   Gary Park, "Canada Reinforces Its Disputed Claims in the Arctic," *Petroleum News* (July 11, 2004). Archived at http://www.energybulletin.net/993.html

249   Clifford Krauss, "Canada Reinforces Its Disputed Claims in the Arctic," *The New York Times,* (August 29, 2004).

250   David Rennie, "Hawks Tell Bush How to Win the War on Terror," *The Telegraph,* (December 31, 2003). Archived at http://www.commondreams.org/headlines03/1231-01.htm

251   Ambrose Evans Pritchard, "China Backs Euro at Dollar's Expense," *The Telegraph* (July 1, 2004)

252   Chalmers Johnson, "Sailing Toward a Storm in China," *The Los Angeles Times,* (June 15, 2004). Archived at http://www.commondreams.org/views04/0715-04.htm

253   Extracted from Bradford C. Snell, *American Ground Transport: A Proposal for Restructuring the Automobile, Truck, Bus, and Rail Industries.* Report presented to the Committee of the Judiciary, Subcommittee on Antitrust and Monopoly, United States Senate, (February 26, 1974), United States Government Printing Office, Washington, 1974, pp. 16-24

254   Tim Starks, "Ridge Reports Investments in Homeland Security Contractors," *Constitutional Quarterly* (September 22, 2004). Archived at http://www.cq.com/corp/show.do?page=crawford/20040923_homeland

255 Michael Lelyveld, "Caspian: Russia Proposes Wider Offshore Zone for Dividing the Sea," *Radio Free Europe*, (August 8, 2002). Archived at http://www.rferl.org/features/2002/08/28082002143233.asp

256 Vladimir Isachenkov, "Russia Plans Nuclear War Games," *Associated Press*, (January 31, 2004). Archived at http://www.miami.com/mld/miamiherald/news/world/7840497.htm

257 "Russia Plans to Double Growth by 2012," *Forbes*, (August 12, 2004). Archived at http://www.forbes.com

258 "Putin: Russia to Increase Military Orders by 40 Percent in 2005," *Channel News Asia,* (August 12, 2004). Archived at http://www.channelnewsasia.com

259 Ambrose Evans Pritchard, "Riyadh Scraps Foreign Reserve Policy," *The Washington Times,* (May 15, 2004). Archived at http://www.washtimes.com/world/20040515-111510-7459r.htm

260 See generally: Richard Heinberg, "Smoking Gun: The CIA's Interest in Peak Oil," *Museletter* (August 2003); Peter Schweizer, *Victory: The Regan Administration's Secret Strategy that Hastened the Collapse of the Soviet Union*, (Grove/Atlantic, 1996)

261 John Pomfret, "Beijing and Moscow to Sign Pact, Stronger Ties Sought to Check US Influence," *The Washington Post* (January 13, 2001).

262 Caroline Gluck, "North Korea Embraces the Euro," *BBC News,* (December 1, 2002). Archived at http://news.bbc.co.uk/1/hi/world/asia-pacific/2531833.stm

263 "Japan and Iran Sign Key Oil Deal," *CNN News,* (February 18, 2004). Archived at http://edition.cnn.com/2004/BUSINESS/02/18/japan.iran.oil.ap/

264 "Selective Service Eyes Women's Draft," *The Seattle Post-Intelligencer* (May 1, 2004). Archived at http://www.fromthewilderness.com/free/ww3/050304_women_draft.html

265 Archived at http://thomas.loc.gov/cgi-bin/query/z?c107:H.R.3598:

266 Archived at http://www.sss.gov/perfplan_fy2004.html

267 David Wiggins, "Draft Creep," *LewRockwell.com* (January 9, 2004). Archived at http://www.lewrockwell.com/wiggins/wiggins7.html

[268] Ibid.

[269] "US, Canada Sign Smart Border Declaration," *CNN* (December 13, 2001). Archived at http://www.cnn.com/2001/US/12/12/rec.canada.border/

[270] "Bush and Kerry, All that Different?" *The Daily Collegian,* (February 19, 2004). Archived at http://www.dailycollegian.com/vnews/display.v/ART/2004/02/19/403425d09cc2a

[271] "US War Machine Nearly Fell Apart, Army Reveals," *Sydney Morning Herald*, (February 4, 2004). Archived at http://www.smh.com.au/articles/2004/02/03/1075776064461.html

[272] Dr. Helen Caldicott, *The New Nuclear Danger,* p.11-12, citing Bruce G. Blair, Harold A. Feiveson, and Frank N.Von Hippel, "Taking Nuclear Weapons off Hair Trigger Alert," *Scientific American* (November 1999).

[273] Ibid.

[274] Ibid at p. 22, citing James Risen, "Computer Ills Meant US Couldn't Read Its Spy Photos," *The New York Times,* (April 12, 2000).

[275] Michael P. Ghiglieri, The Dark Side of Man: Tracing the Origins of Male Violence, p 190 as cited by Jay Hanson, "A Means of Control" *Brainfood* (January 1, 2001). Archived at http://www.dieoff.com/page185.htm

[276] Major Ralph Peters, "Constant Conflict," *Parameters*, (Summer 1997), p. 4-14. Archived at http://www.informationclearinghouse.info/article3011.html

[277] Linda Diebel, "Bush Doctrinaires: Analysts Point to Strong Signs America's War Machine Will Continue to Roll," *Toronto Star* (April 13, 2003). Archived at http://www.commondreams.org/headlines03/0413-06.htm

[278] "Beware Mad Max World of US," *Sydney Morning Herald*, (October 15, 2003). Archived at http://www.smh.com.au/articles/2003/10/15/1065917483266.html

[279] The similarities between Peak Oil and dehydration were brought to my attention by a reader of my site who, while serving as a US Army Ranger, almost died from dehydration.

[280] David Price, "Energy and Human Evolution," *Population and Environment: A Journal of Interdisciplinary Studies,* Volume 16, No. 4. p 301-319, (March 1995). Archived at http://www.dieoff.com/page137.htm#2